NANO
NATURE

NANO NATURE

Richard Jones

METRO BOOKS
NEW YORK

This 2008 edition published by Metro Books,
by arrangement with HarperCollins Publishers Ltd.

Editors: Hugh Brazier and Julia Koppitz
Proofreader: Janet McCann
Cover design and design layout: Emma Jern
Design: Marc Marazzi
Index: Ben Murphy
Production: Keeley Everitt

Metro Books
122 Fifth Avenue
New York, NY 10011

ISBN-13: 978-1-4351-1033-5

Printed and bound in China

10 9 8 7 6 5 4 3 2 1

CONTENTS

Size matters. For both the very large and the very small organism, size is a major determinant of body shape, body composition and body structure. In large plants and animals it's all a matter of how strong the supportive edifice of woody trunk or skeletal bones has to be to overcome the power of gravity pulling down, and at the same time how much energy is required to push nutrients to the outer extremities of leaf or limb. Elephants can only grow so large by having relatively huge leg bones to carry them about, and redwoods can only grow so high because there is a limit to how far water from the roots can reach up the trunk.

Humans are denizens of this larger world. Even the greatest of all plants and animals on the planet are not really that much bigger than us. We can see them clearly, we can touch them, feel them, measure them, use them even – we are intimately familiar with them.

At the other end of the size spectrum, gravity reaches a point of almost insignificance for tiny organisms. How else could houseflies walk upside down on the ceiling? Moving oxygen, carbon dioxide, nutrients and waste materials in, about and out of the body is a short easy journey. There are now different, strange, alien constraints on the growth and form of much smaller creatures, and smaller structures.

Unaided, the human eye can only see so far into this small world. But even the humble hand-lens opens up a marvellous hidden world – bizarre and beautiful – and a light microscope shows even greater wonder, as clarity of vision is brought to bear on individual bristles on individual insect limbs. Now, at the far end of human technological achievement, the scanning electron microscope opens up a stunning world of improbable structures and fanciful patterns. Any familiarity is lost. This is altogether another world – the nano world.

In the nano world, growth, form and structure are controlled at the level of individual cells or even individual molecules. Here there are processes going on at the very edge of human understanding. Just as the first microscopes revealed a microcosm in a drop of water, or the microstructure of a bee's knees, the scanning electron microscope shows unimaginable detail and unexpected formations.

More surreal than any human art, more complicated than anything fashioned by human hand or machine, these images from the high-tech science laboratory are brought into the public domain in *Nano Nature*. These glimpses are as striking and ground-breaking today as they were over 150 years ago, when the first close-up coloured images of coral polyps, insect wings and red blood cells, as viewed through a light microscope, so thrilled our Victorian ancestors.

Scanning electron microscopy

Unlike conventional microscopes, which use glass lenses to focus light on an object being viewed, a scanning electron microscope (SEM) uses magnetic wire coils to focus a narrow beam of electrons across a target. The reflections of some of these electrons hit a sensitive screen, where they are converted into electrical signals that can be analysed by a computer to generate the image. To increase the reflectivity and to give better contrast in the image, the object first has to be preserved with solvents and stabilised by freezing and drying. It then has to be coated with a minute layer of metal, usually gold or platinum, just a few atoms thick. This is achieved by a process called sputtering, where an electrical current is passed through the metal electrode until it is incandescent and a stream of atoms sprays off from the tip.

Unlike light, which humans (and other animals) see in different colours, SEM images are monochrome. And striking though these may be, they can be greatly enhanced by the addition of false colour. This process, analogous to the artistic hand-tinting of sepia photographs when cameras were first invented, is not intended to corrupt or distort, or to create a false impression of what the SEM is viewing. Rather it enriches the pictures, producing the exquisite images we can now see.

About this book

Nano Nature is divided into two sections, on the themes of 'nano form' and 'nano function'. In the first section of the book, the abstract beauty of the close-up images provides insights into different perspectives of form. The scrutiny of an object is lifted from mere clinical examination to an appreciation of the wonderful artistry of nature. This is brought about by one of the most astonishing aspects of scanning electron microscopy – the crispness of the images and the great depth of the focus. These produce the most startling views, and give a truly three-dimensional appearance. This is particularly striking for objects that might normally be assumed to be flat – a butterfly wing, a leaf surface, shark skin. At the extraordinarily high magnifications now available to us, these are revealed as rows of tombstones, forests of hairs, or rivers of teeth.

In the second section, the tools, the weapons and the sense organs of plants and animals are examined from the point of view of their detailed function. These are the precise mechanisms of life by which plants and animals feed, grow and reproduce. Again, the astounding clarity of the images shows so much in the fine detail. A bee sting is no mere prick of a pin, it is a murderous-looking shaft, barbed, honed and full of menace. The breathing pore on the surface of a leaf is positively pouting, pursed, ready to tell us its secrets.

The images presented here have been chosen primarily for their dramatic impact. Each is a work of art in its own right. But woven through the text are the themes of nature, the themes of biology.

The prime directive of any organism is to reproduce, to create offspring, to pass on genes to the next generation – so mechanisms of mating, egg-laying and seed production are well represented. To achieve this genetic immortality, organisms also need to eat and breathe, and avoid being eaten or killed, long enough to reproduce – so a variety of strategies for feeding and respiring, and for defence against predators and a hostile environment, also features in these pages.

A book such as this could never be a comprehensive guide to nature, but, like the scanning electron microscope itself, it offers a selection of vivid pinpoint perspectives, tasters of life, snapshot portraits of a wondrous world in miniature.

Measuring the nano world

The standard unit of measure adopted by the scientific (and much of the non-scientific) world is the metre. This was first defined by the French government in 1791 as one ten-millionth of the length of the meridian line passing through Paris from north pole to equator. Or rather it was an extrapolation – because the only part of the globe measured was between Dunkirk and Barcelona. This rather arbitrary measure was provisionally set in brass in 1795, then officially in platinum in 1799. That prototype, the Mètre des Archives, still exists. Later it was copied and international prototypes were established.

Metal rods, however, expand and contract according to their temperature, and the original calculations did not take into account the flattening of the earth due to its rotation.

Today a metre is defined as the length of the path travelled by light in vacuum during a time interval of 1/299,792,458 of a second.

- 1 metre (m) = 1,000 millimetres (mm)
- 1 millimetre = 1,000 micrometres (μm)
- 1 micrometre (sometimes called a micron) = 1,000 nanometres (nm)

1

Nature is full of beauty. A grand vista of snowcapped mountains or a verdant wooded valley is available for all to see and wonder at, but much is invisible to the naked eye. The scientific approach to the natural world no longer sees this beauty as evidence of the art of the Creator, but natural form continues to fascinate, and there is as much beauty in a micrograph of a fly's eye, or a slug's tongue, as in the intricate pattern on the wing of a butterfly. In this section the details revealed by the scanning electron microscope are treated as abstract images, many of which would not look out of place in an art gallery. The images are grouped primarily by form.

NANO

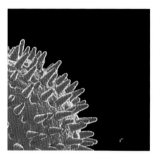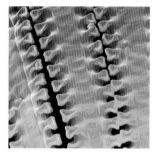

single cells · scales · hairs · tongues
folds · pollen and spores · leaves · domes
surface extrusions

FORM

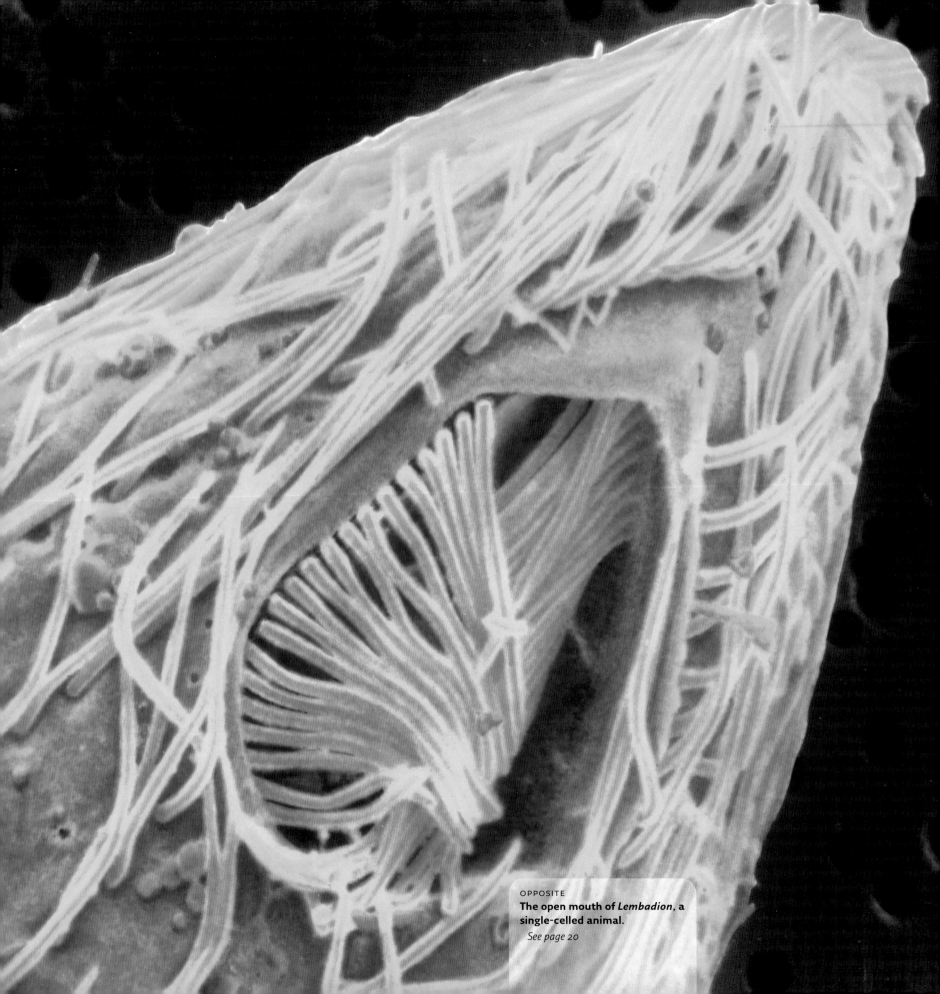

OPPOSITE
**The open mouth of *Lembadion*, a
single-celled animal.**
See page 20

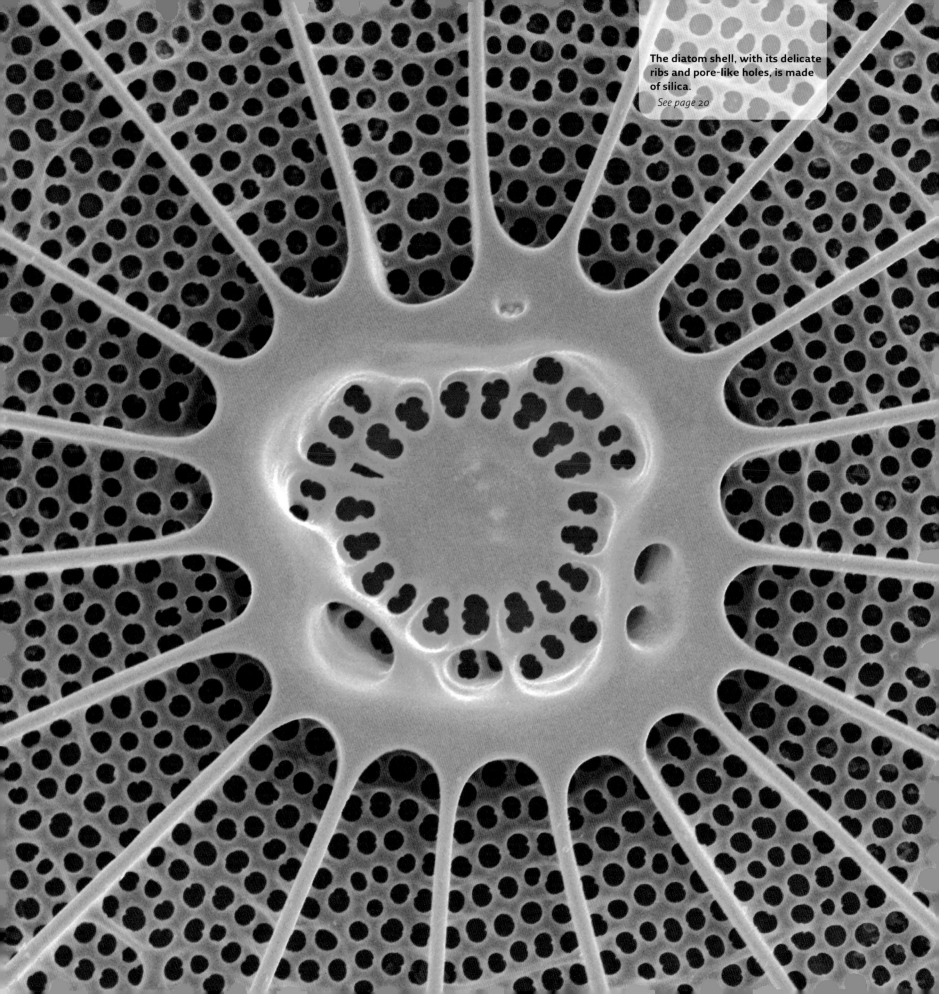

The diatom shell, with its delicate ribs and pore-like holes, is made of silica.
See page 20

OPPOSITE
The internal skeleton of a radiolarian, a single-celled oceanic animal.
See page 21

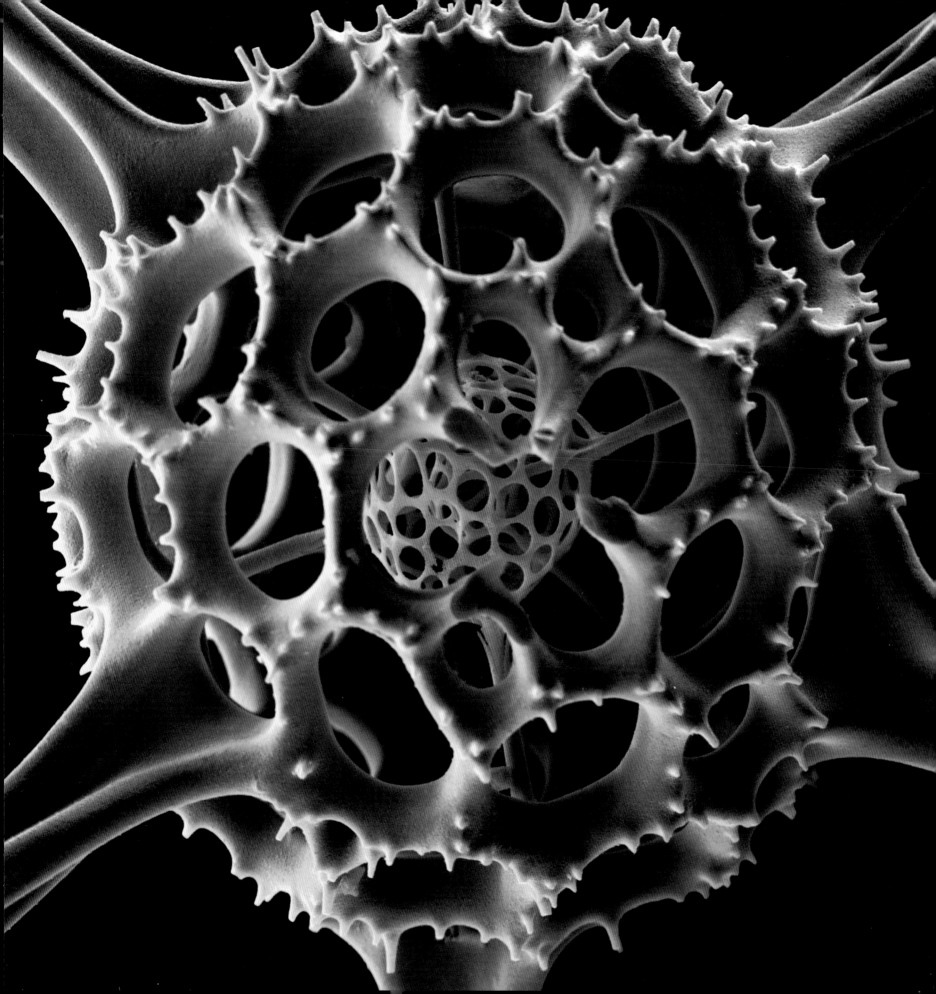

OPPOSITE
Giant chromosomes from the maggot of a fruit fly, *Drosophila*.
See page 21

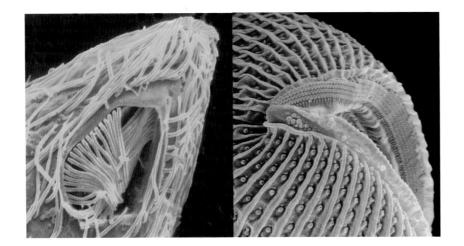

Resembling a plump knitted or pleated coin-purse, mouth unzipped and open, ready to receive cash, the exterior of *Lembadion bullinum* is very much a bag-like structure, but the opening is a real mouth, and it is ready to devour prey. *Lembadion* is a single-celled animal (a protozoan) found in freshwater lakes, ponds and slow-moving streams. It is oval or kidney-shaped, with a single large mouth opening called a cytostome (coloured green here) on its underside. It moves about by rhythmically whipping minute hairs, called cilia, which cover most of its body. In the process of preparing this specimen for micrography the cilia have been removed, but their arrangement can still be seen in the pleated rows of pores, called kinetosomes, where each hair would have projected. *Lembadion* has about 50–65 rows of kinetosomes, with 40–55 cilia sprouting from each row.

The cilia beat in unison, or in waves,

to move *Lembadion* through the water. The creature is thus able to pursue other minute single-celled animalcules, as well as bacteria and algae, on which it feeds. The beating hairs also waft food items towards the open mouth. Fine hairs around the buccal (mouth) opening fold over to push food in like a giant swallowing mechanism.

Most individuals of *Lembadion* are only 100–120 μm long, but giants of 200 μm long have been reported when food becomes scarce. There are in fact several intermediate-sized forms, each adapted to feed on prey items of different sizes.

Most protozoans reproduce by binary fission, with the body simply dividing to create two daughter cells. By delaying this division during the most extreme food shortages, *Lembadion* is able to grow up to twice its normal size. However, when food again becomes more readily available, rapid cell divisions bring the body size back down to normal.

Delicate spindle spokes radiating from a central hub are an ancient wheel design. But this one is far older than the first wheel invented by our ancestors, for it evolved hundreds of millions of years ago. Diatoms are minute marine algae, each consisting of a single plant cell coated with a casing of extruded silica (silicon dioxide), called a frustule. The largest known species is 2 mm across, but most are microscopic. There are estimated to be over 100,000 species of diatom in the world's oceans, with hugely diverse shapes, forms and ornate patterns.

In spring and early summer, when nutrient levels in the upper water are high, diatoms reproduce quickly and soon come to dominate the community of phytoplankton (single-celled algae).

Each frustule is made of two shells or valves, like two wide flat cake-tins, one slightly larger and overlapping the other. During the boom phase of the reproductive cycle a diatom splits to form two daughter cells. Each new cell

inherits one of the original valves and then secretes a second slightly smaller valve within it. With further splits the average diameter of the diatoms becomes smaller, until they reach a minimum size beyond which a cell cannot sustain itself. At this point, usually coinciding with a fall in oceanic nutrients, the diatoms produce much larger spores, and these act as a resting stage until the seasonal currents bring new nutrients the following year.

The unique shell of the diatom is created inside the organism by polymerisation (the sequential joining together) of individual silicic acid molecules into long silicate chains, which are then extruded to the outside of the cell and added to the cell wall. Nanotechnologists are now studying the genetics and self-replicating ability of diatoms in the hope that an understanding of the well-ordered and microscopically precise mechanisms of silicate deposition may help in the manufacture of silicon chips.

RIGHT
The rows of small pores show where cilia were attached to the *Lembadion*.

LEFT
The cilia of a close relative, *Tetrahymena*, are still in place, and the buccal cilia around the mouth are in the 'swallow' position.

RIGHT
The delicate ribs and pore-like holes (called striae) of the diatom shell are made of silica, giving beautiful form and structure to the single plant cell inside.

LEFT
The simple yet beautiful shapes of diatoms are hugely diverse, with an estimated 100,000 different species known.

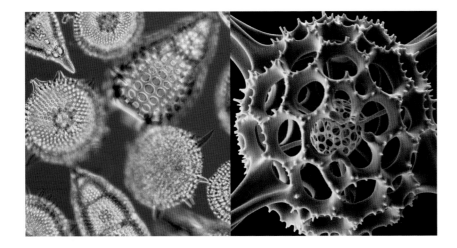

This spiny-faceted hollow golf-ball shape is the empty skeleton of a radiolarian, and like vertebrate skeletons it has the dual function of protecting the body and maintaining its shape. Radiolarians are tiny single-celled animals that form part of the oceanic plankton. They range in size from a few hundredths to a few tenths of a millimetre across. Although their soft gelatinous bodies have the texture of an amoeba, they maintain a regular shape because of a complex internal skeleton made of silica, which is secreted into many bizarre and beautiful shapes.

The inner chamber of the skeleton, surrounded by a membrane, contains the endoplasm, the semi-liquid cell substrate, and cell organelles such as the nucleus (with the organism's DNA), mitochondria (energy-producing bodies) and the Golgi apparatus (responsible for protein and fat secretion). Outside the shell is the ectoplasm (also called the calymma), a similar protein gel, which covers the projecting spines and which often contains tiny single-celled algae living inside it. These symbiotic algae help their hosts by harnessing the energy of sunlight through photosynthesis, and in return receive shelter and an endless supply of carbon dioxide from the radiolarian.

The ectoplasm is also used for feeding. Blob-like extensions called pseudopodia grow out and surround food particles, which then become absorbed into the cell body, where they are digested. In laboratories a radiolarian can also manoeuvre itself about inside a glass vessel using the pseudopodia like tentacles, but they have not been found attached to anything in nature.

Because of their stony skeletons, radiolarians are relatively well preserved in the fossil record, and they have been evolving a huge range of intricate shapes for over 600 million years. Despite their variety, there are two basic structural themes: the spherical-shelled Spumellaria (shown here) and the conical-shelled Nassellaria.

No matter how hard we look, the beads of this necklace are not quite crisply defined, their boundaries merging and distorting. But then these are multiple chains of DNA, and the bands of colour are the nearest we get to a visual image of how genes are laid out in a living organism.

Chromosomes are aggregations of DNA (the genetic material) in the nucleus of animal and plant cells. They are usually only visible (under high-power microscopes) during cell division, when the DNA condenses and contracts from a homogeneous translucent blob into a series of distinct sausage-shaped entities – the chromosomes. This coincides with a duplication of the DNA, and therefore also a doubling of the number of chromosomes; the chromosomes are pulled apart to form two equal daughter cells.

Chromosomes are ordinarily very small, but in 1881 the French embryologist Edouard-Gérard Balbiani found greatly enlarged examples in the salivary glands of midge maggots. They have since been found in the larvae of many different fly species, including the fruit fly *Drosophila melanogaster*. In 1935 the banding of *Drosophila* giant chromosomes was sketched by American geneticist Calvin Bridges, and these diagrams are still used today as maps of the fly's genetic structure.

Giant chromosomes (also called polytene chromosomes) are formed by repeated duplication of the DNA strands, but without any separation of daughter chromosomes and without any cell division. Instead, the cell containing the chromosomes grows much larger than normal. The multiple copies of the DNA, up to 2000 of them, remain aligned with each other, but 'puffs' show where they are uncoiled when actively manufacturing cell products. It is thought that the fly larvae benefit from giant chromosomes because a cell containing multiple copies of a gene can manufacture proteins at a much faster rate than normal.

RIGHT
The internal skeleton of the radiolarian is made of silica, and gives form to the otherwise shapeless amoeboid protozoan.

LEFT
The glistening perforated jellies of live radiolarians come in many different forms, but are broadly grouped into spherical or conical shapes.

RIGHT
Giant chromosomes from the larva of a *Drosophila* fruit fly, each containing up to 2000 times the DNA found in the chromosomes of normal cells. The bands can be used as a map in genetic studies.

LEFT
A bithorax mutant of *Drosophila*, with a duplicated thoracic segment and four, rather than two, wings.

OPPOSITE
**Aligned scales on the wing
of a butterfly.**
See page 30

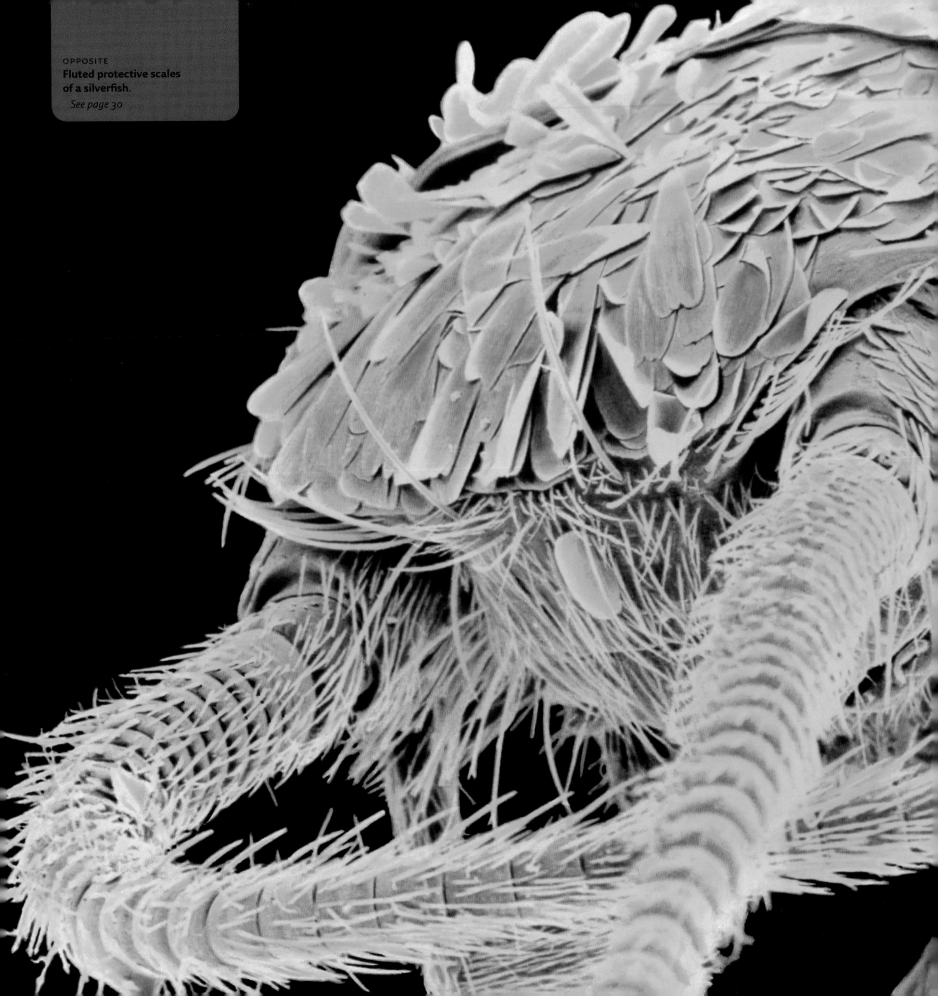

OPPOSITE
**Fluted protective scales
of a silverfish.**
See page 30

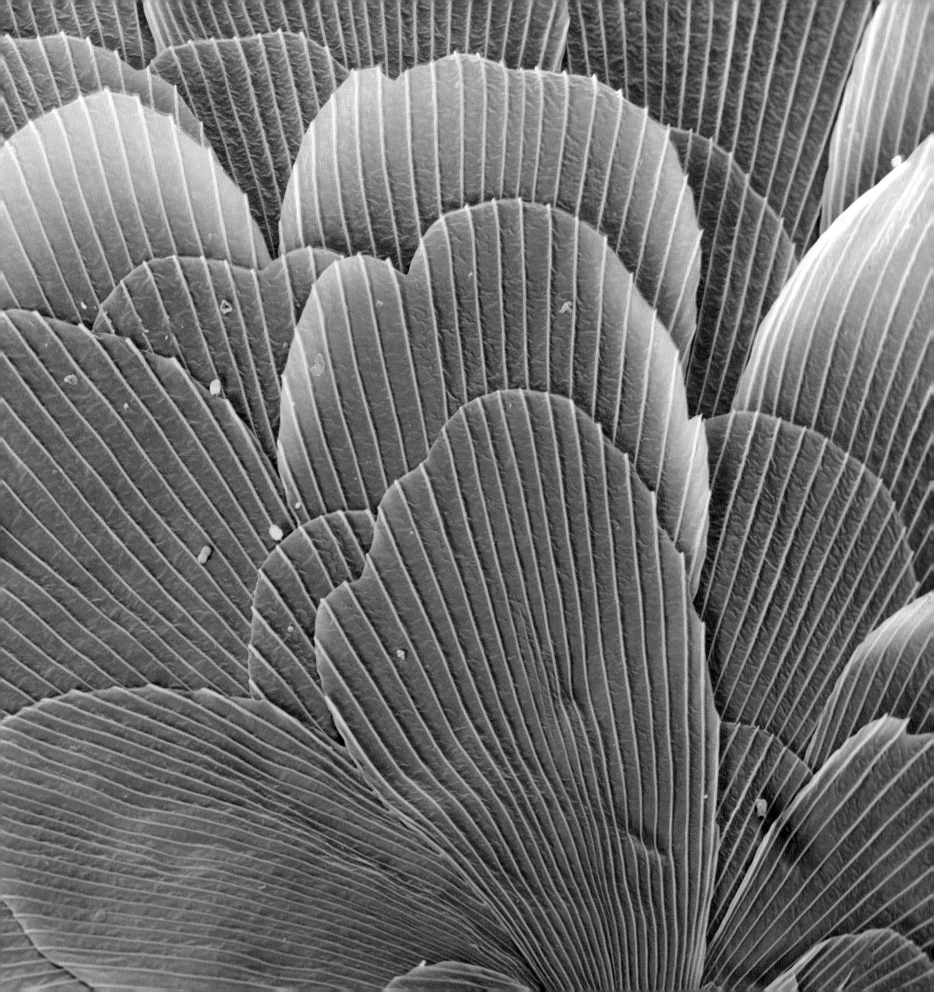

OPPOSITE
**The flattened scales covering a
beetle's body.**
See page 31

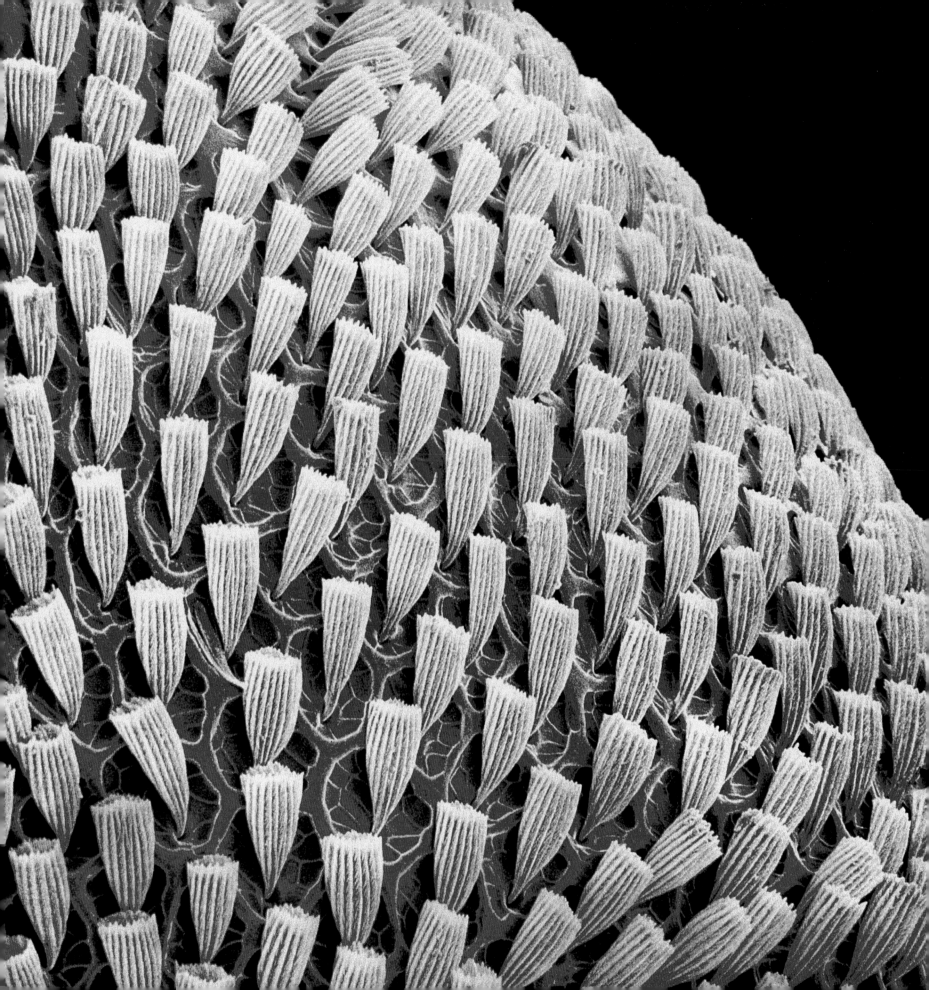

OPPOSITE
**The raised hydrodynamic scales
of a shark's skin.**
See page 31

Arranged like rows of seats in an auditorium, some of the cloth covers faded and peeling loose, these flat plates may look untidy close up – but from a suitable distance they present a pristine picture of the individual wing scales on a butterfly.

Butterflies and moths form the insect order Lepidoptera. The name means scale-winged, referring to the dense and distinctive patterning made up of tiny flakes of colour. The scales are flattened and grossly distorted hairs (called setae or trichia), and they are made from chitin, the same substance that forms the majority of an insect's exoskeleton.

The scales of a butterfly are easily determined from the regular regimented ranks. Moths tend to have a much more random arrangement, even though the patterns can still be precisely formatted across the wing. In both butterflies and moths the colour pattern of the wing is under precise genetic control.

Scales have several functions for butterflies, moths and the many other insects that have them (beetles, silverfish, bugs and caddisflies, for example). The colours they provide can be combined into an infinite variety of mosaic-like patterns. Because of their slanting three-dimensional form, they can also give a glossiness beyond the level of a straightforward pigmented background. They can provide a glinting metallic lustre as dense parallel sub-microscopic grooves and channels refract the light into its rainbow colours, emphasising the blue (as in the magnificent *Morpho* butterflies of tropical forests) or green (as in the huge birdwings of southeast Asia) or gold and silver (as in the silver-Y and golden-Y of Britain and Europe).

Scales can also be safely shed if danger threatens. It is better for a butterfly to lose some of its delicate scales to the beak of a bird, or the sticky threads of a cobweb, or the fingers of a child, rather than to lose its life.

These scalloped and ribbed scales would not look out of place on a medieval knight in armour, or on a fresh salmon laid out on the fishmonger's slab. The reason is simple, for although these delicate overlapping plates are several orders of magnitude smaller, they serve the same purpose – tough but flexible protection of the underlying skin. Silverfish get their name from this close resemblance of their body covering to fish scales, and though they are sleek and metallic they are, in reality, no fish.

Silverfish are small insects, 10–15 mm long, part of the order Thysanura. Members of this order are also called bristletails, because of three long hair-like bristles projecting from their tails. There are about 400 species of thysanurans known from around the world.

Silverfish are part of one of the oldest and most primitive groups of insects, but they have now become almost entirely household creatures. Indoors they feed on spilled foodstuffs like cereals, breadcrumbs and sugar, as well as grazing on household goods such as clothes, carpets, books, wallpaper and glue. Another bristletail, the firebrat, gets its name because it was found in heated buildings, especially bakeries, where it would feed on spilled flour around the ovens. Often the presence of silverfish is only detected because of evidence of their nocturnal feeding forays. During the day they hide away in cracks and crannies, behind skirting boards and loose wallpaper, and under floor coverings.

Silverfish are wingless and relatively short-legged, but they are extremely agile and move with surprising speed back into hiding if they are disturbed. Their silver scales are also a protection against being caught by predators, because they flake off, leaving an enemy with a mouthful or a handful of silver dust, while the silverfish makes its getaway.

RIGHT
In the SEM image the rows of scales are artificially tinted mauve or blue to show their different flat or fluted forms, but in life the colour patterns are virtually infinite.

LEFT
The beautiful pattern of a swallowtail butterfly's wing is made up of a complex mosaic of scales, each one a separate and distinct colour.

RIGHT
Overlapping silver scales give the silverfish its name. As well as offering it armoured protection, they allow agile flexibility.

LEFT
Silverfish are primitive wingless insects with long multi-segmented antennae and chewing mouthparts.

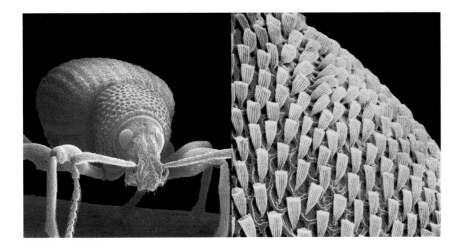

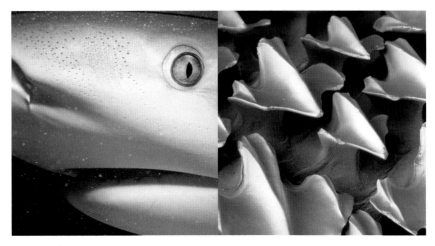

This could be the science-fiction version of a futuristic aerodrome crowded with delta-winged hypersonic jet planes. Each appears lodged in its own docking bay, a sculpted shallow depression in the surface of the unworldly planetoid. In fact each is a minute scale, covering the back of a very earthly beetle.

Some insects, notably butterflies and moths, but also very many beetles, are covered in tiny flattened scales. Like the whole of their hard body shells, the scales are made of a tough waterproof substance called chitin. This polysaccharide molecule, derived from carbohydrate sugars, is laid out in layered sheets giving great strength despite their lightness – just as plywood, made from thin sheets of wood, is tough and flexible.

Insect scales are evolved from chitinous hairs, which grow from dimpled pores all over the body. In some insects these setae have developed into sense organs of touch, or they detect sound vibrations and air movements. In others they have become spines used in defence or attack. Where they have developed into scales, their flattened forms allow them to act like tiny mosaic tiles, producing diverse colours and patterns.

Each scale has a genetically coded colour. This is usually based on the pigment melanin, which is found throughout the animal kingdom. In different densities it forms a colour spectrum from black through brown, red, orange, yellow and white. In addition, metallic colours are created by sub-microscopic grooves that split the light and reflect back a shining blue, green or bronze.

The mosaic patterns of the scales can be aligned to create bars or blocks or stripes of colour. These scales are on the back of a hide beetle, which feeds on carrion, and also on stored animal skins and furs. Its scales form a variegated stippled mix, to disguise it against the mottled fur or feather of its food.

These sharp and jagged teeth look set to gouge whatever they touch, and could be the chiselled blades of a rotary coal-mining auger or earth-digging machine – or some giant cheese grater. Each three-pointed node is raised away from the surface, as if to allow the chaff and rubble to escape. In fact, these honed plates are the tiny scales on the skin of a shark, and the only cutting they do is through the water.

The resemblance to teeth is no accident, however, because they are thought to have evolved in the same way, and share many structural details. Called placoid (plate-like) scales, they are composed of dentine, a tough calcareous material denser than bone; this is made up of a hard crystalline substance called apatite embedded in a flexible protein, collagen, which is a major component of skin. The top is coated with a smooth enamel, and each is supplied from beneath by a capillary blood vessel. They are so similar to teeth that most modern sources use the term dermal denticle (small skin tooth).

The denticles provide a tough but flexible chain-mail coat of armour, preventing barnacles and parasites from attaching themselves, and there are even reports of sharks using them to inflict damage on prey. But perhaps their most useful quality is giving the shark silent and streamlined stealth as it moves through the water.

At first sight the sharp pointed ends of the denticles appear to be the front, but these are really the trailing edges, and the shark in this image has its head to the top left. The series of three ridges and two grooves on each scale is a hydrodynamic adaptation to reduce drag as the fish swims. By guiding the water past the shark skin, turbulence is reduced and eddies are prevented, effectively silencing the sound of the water passing over the shark's body so that its fast approach to prey is not noisily advertised.

RIGHT
Each fluted triangular plate is a scale on the back of a beetle shell. Each scale is a particular colour, and together they can create a wide range of patterns.

LEFT
Camouflage is especially important for beetles which, like this weevil, are exposed while feeding on leaves.

RIGHT
Tough enough to graze wood and metal, shark skin has long been used as sandpaper.

LEFT
Sharks are supreme marine predators, and their sleek shape gives them speed and manoeuvrability in the water.

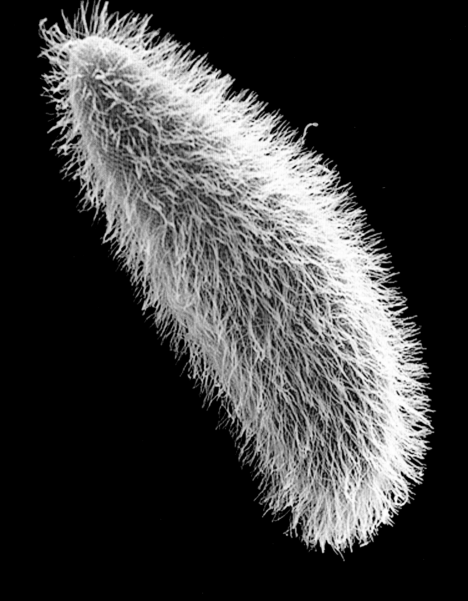

OPPOSITE
Cilia (moveable hairs) covering the body of a tiny single-celled animal.
See page 40

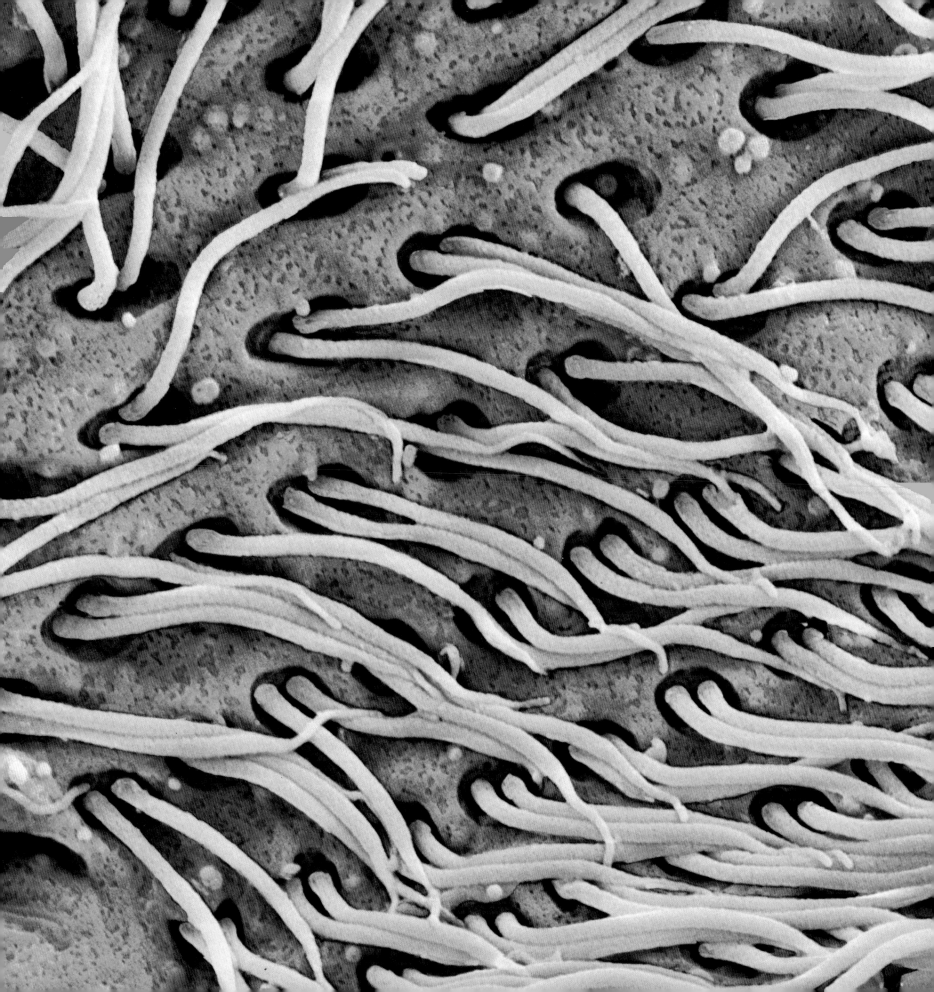

OPPOSITE
Dead cells covering the smooth surface of a cat's whisker.
See page 40

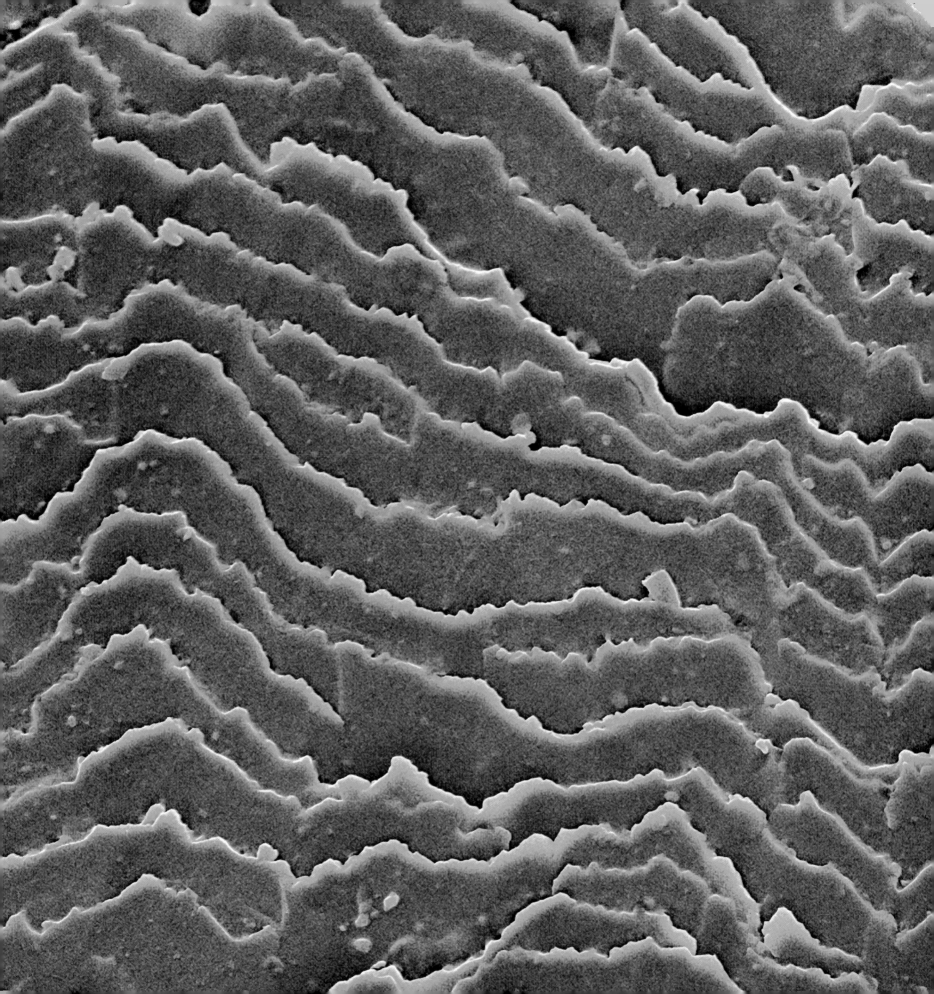

OPPOSITE
Pad of spatula-shaped hairs on the foot of a fly.
See page 41

OPPOSITE
**Pads of splayed hairs on
a gecko foot.**
See page 41

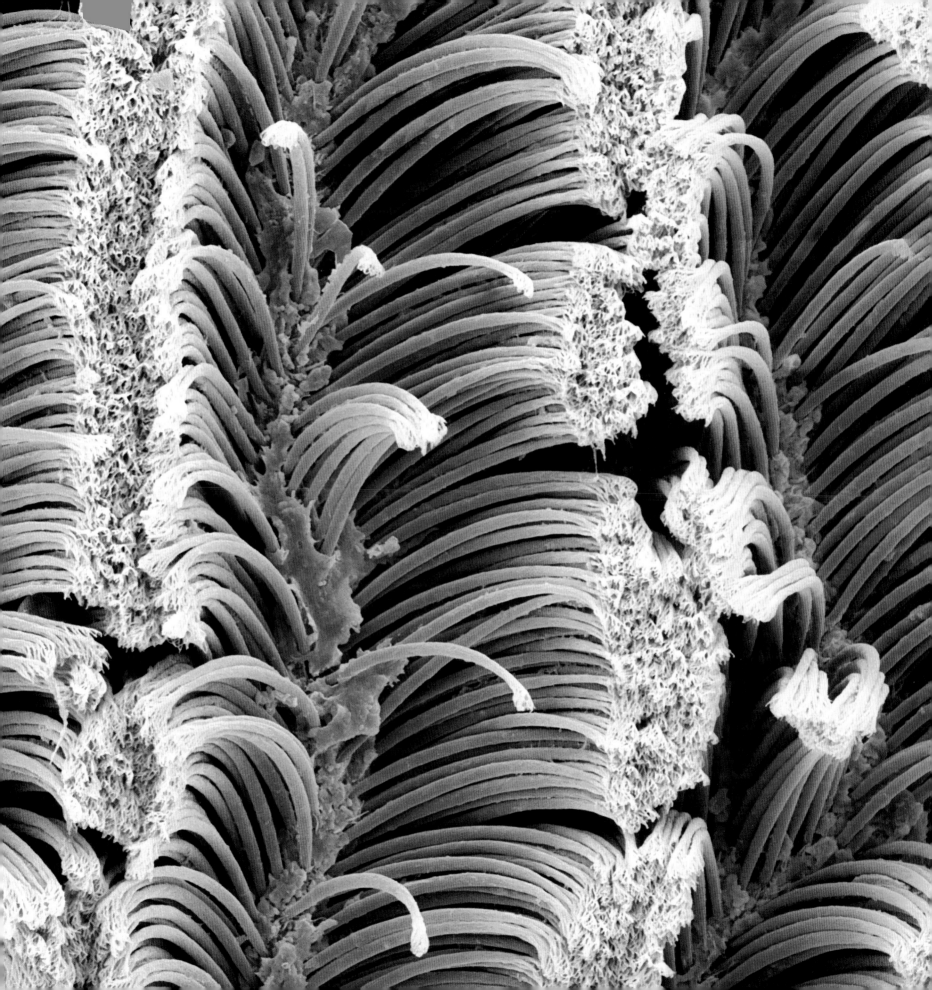

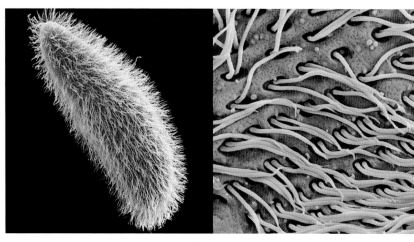

The fragile hairs projecting through the outer skin of this microscopic single-celled animal now look limp and unkempt, but in life they were stiff and erect and orchestrated like banks of tiny oars to power it through the water. Ciliates take their name from this covering of minute hairs, or cilia (singular cilium). They are the most important group of single-celled animals, with about 10,000 known species, and are found wherever there is water: in oceans, rivers, lakes, ponds, puddles, soil.

The cilia beat rhythmically, in unison or in waves, and propel the minute animal through the water. They are usually arranged in rows, with each pore, or kinetosome, giving rise to either a single hair or (as here) double hairs.

The motion of an individual cilium has been likened to a whiplash, but recent advances in microscopy have revealed it to be more like the beat of a wing. In the downward (power) stroke the cilium remains relatively straight and stiff, pushing the water backwards. In the forwards (recovery) stroke it flexes to drag back through the water. The effect of many hundreds or thousands of cilia on the protozoan's body is to pull the animal through the water.

The precise mechanics of cilium action are still not completely understood, but the amount of bending along the hair is controlled by minute tubes of special proteins that run along inside. Each cilium has a core of two microtubules surrounded by nine pairs of microtubules. The controlled sliding movement of the microtubule pairs allows the hair to curve at different points along its length.

Cilia, also called flagella (singular flagellum), are remarkably similar throughout the animal kingdom. In larger animals they are usually attached to the outside of a stationary cell, and their wafting action moves, for example, dirt and mucous out of the lungs along the trachea, or an egg along the Fallopian tube to the womb.

The ragged edges of these uneven overlapping plates suggest some tough coarse use, like the damaged scales of a fish whose scaly armour has protected it from attack. But the abraded appearance under high magnification is very far from their sleek smoothness in life, because this is a cat's whisker, the epitome of neatness and precision.

The resemblance to fish scales is an apt one, because hairs are specially adapted from skin, and are covered in the remains of tightly packed epidermal cells. These cells create the main component of the hair – the long spiral chain molecules of a protein called keratin. This natural polymer gives hair its strength and its flexibility.

Whiskers, technically called vibrissae, are many times longer, thicker and stiffer than ordinary hairs, and have become important sense organs for some animals. In cats they are usually arranged in four or five rows on the mystacial (moustache) pads above the upper lip. There are also some shorter whiskers above the eyes. Each vibrissa is much more deeply embedded in the skin than ordinary hair follicles, and its base is surrounded by a capsule of blood, called a blood sinus, which itself is surrounded by nerves. Even the tiniest movement at the tip of a whisker bends the root, altering the pressure at different points around the blood sinus – and this is detected by the many nerve cells that penetrate it.

Cats use their whiskers to feel by touch, especially in darkness, but also to detect tiny movements and changes in air currents. It is a myth that a cat's whiskers are precisely as long as its body is wide, allowing it to judge whether or not it can squeeze through a narrow gap. Whisker length is under genetic control and varies from breed to breed, but cats certainly use them to feel their way about.

Cats also use their whiskers as a means of communication, drawing them back and down if they are angry or alarmed, pointing them forwards and up if they are alert and curious.

RIGHT
Cilia are minute hairs that cover the bodies of single-celled animals and move like oars to beat in unison, pulling the creature along.

LEFT
Paramecium, a freshwater ciliate protozoan, uses its thousands of beating cilia to pursue bacteria and other small prey.

RIGHT
The remains of dead skin cells cover a cat's whisker, which, like other mammalian hairs, is derived from the epidermis.

LEFT
At the root of each whisker is a blood-filled cavity surrounded by nerves which detect any movement in the stout hair.

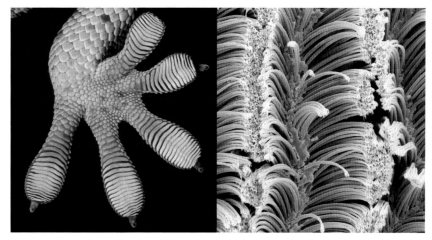

A garden catalogue would do well offering such an intricately delicate ornamental thistle or teasel flower. What appear to be the soft curled petals of the flower contrast with the sharp spines of the stem, creating an image both striking and attractive. But this is not part of the plant kingdom, it is the tip of a fly's foot, and its attractiveness is in the power of molecular adhesion.

A fly's tarsus (foot) is divided into four or five hinged segments, each called a tarsomere, and under the terminal tarsomere is a cushion-shaped pad called a pulvillus. The pulvillus is covered with hundreds of tiny spatula-shaped hairs known as tenent setae. These flattened hairs are the key to the fly's ability to walk upside down on the ceiling, or cling on to even highly polished glass windows. Precisely how these work is still being studied. Flash-freezing of fly footprints on glass shows that a liquid is used, but 'dry' adhesion by van der Waals

forces (as used by the gecko) may also be involved.

When the fly lands on a smooth surface, the pulvillus pad is pressed down hard and the tip of each tenent hair exudes a minute amount of liquid. It is this liquid which holds the fly upside down on the ceiling, or glued to the glass. The force keeping the insect suspended is not friction, but a molecular pull similar to surface tension, much like that which keeps droplets of water suspended from the bathroom ceiling. It is produced by individual molecules in the liquid attracting each other, and at the same time attracting molecules in the tip of the tenent setae and in the fabric of the ceiling.

Although powerful, this force can still be usurped. Carnivorous pitcher plants coat their water-trap containers with an oily wax that disperses the fly's secretions: the insect cannot maintain a grip, and it slips and slides to its doom.

The tufted fibres of what appears to be a deep shag-pile carpet look slightly worn and faded at their tips. But this is not because too many feet have walked across them. Instead, this is the hairy underside of a foot that does the walking. Most dramatically, though, this foot does most of its walking up the wall and across the ceiling, rather than on the floor. It is a gecko's foot, and its hairy feet give the gecko the ability to cling to the shiniest of leaves in the forest, and the glossiest of glass in the laboratory.

The ability of geckos to cling on tight to flat surfaces has fascinated observers for millennia. A gecko can hang from a ceiling by a single toe, but it uses more than just claws to grip. Various ideas have been put forward, including friction and capillary attraction (based on the surface tension of liquids), but the exact mechanism has only recently come close

to a solution.

Each gecko foot has half a million tiny hairs, called setae, 30–130 μm long, and each hair terminates in many hundreds of even tinier projections of 0.2–0.5 μm. When the gecko places its foot against a smooth surface, there is so much touching connectivity that forces which usually only attract molecules to each other (called van der Waals forces after the Dutch scientist who predicted them) are able to keep the animal suspended.

Each seta produces only a minuscule force, but together they create such a binding attraction with whatever they are touching that it takes ten times the gecko's body weight to prise the animal off. This dry adhesion (without liquid or sticky glue), which can be stuck, peeled off and stuck again, many times, is already tempting manufacturers to come up with a gecko tape for everyday use.

RIGHT
The tiny flattened hairs on the pulvilli allow the fly to cling to polished surfaces such as glass by secreting droplets of liquid.

LEFT
A flesh fly hangs from a pane of glass using the broad fan-shaped pads (pulvilli) on its feet.

RIGHT
Many hundreds of thousands of small hairs on the soles of its feet allow a gecko to walk on smooth surfaces by harnessing the power of molecular attraction.

LEFT
Pressed against a sheet of glass, the hairy pads are splayed out to give maximum grip.

OPPOSITE
Finger-shaped grooming lobes on a rabbit tongue.
See page 50

OPPOSITE
The ribbon of sharp backward-pointing teeth on a slug tongue.
See page 50

OPPOSITE
**The complicated sieve channels
at the tip of a fly's proboscis.**
See page 51

OPPOSITE
Taste buds and rough-gripping surface of a bat's tongue.
See page 51

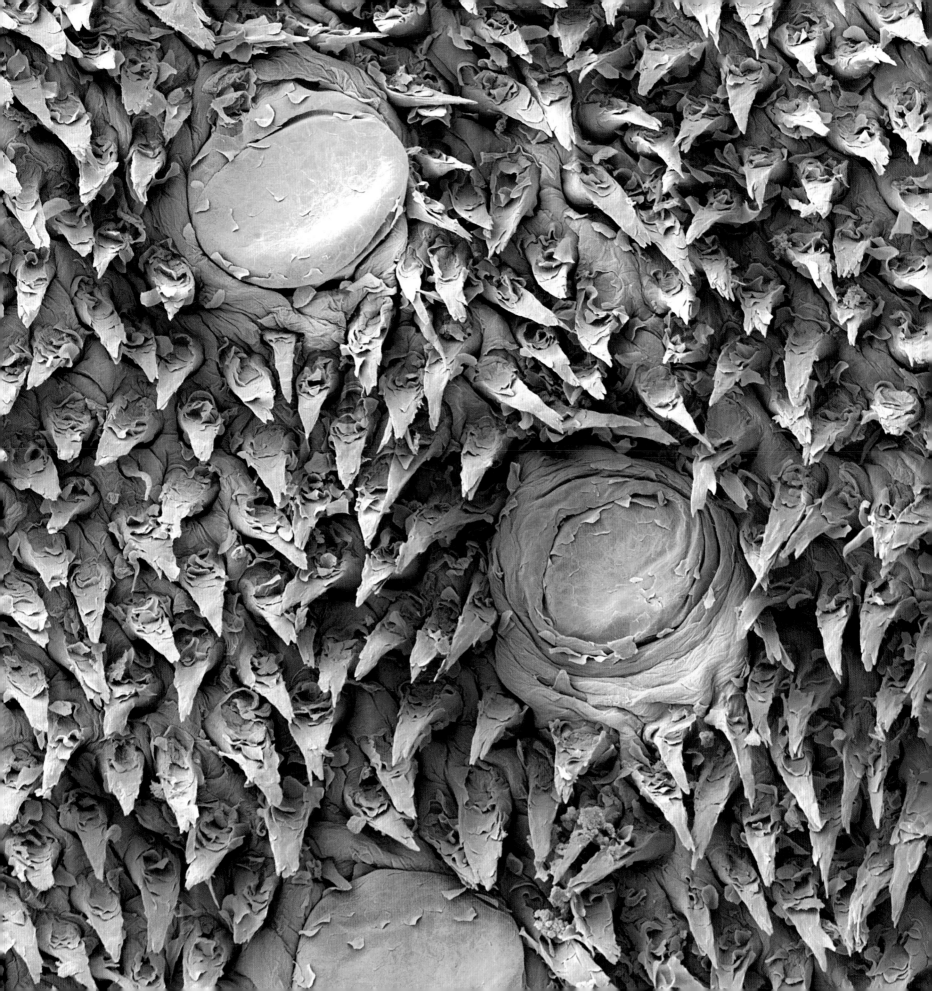

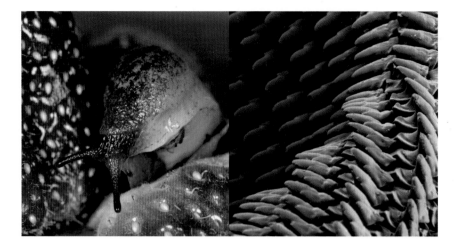

These fingered scaly cushions are arranged in vague diagonal rows. They are not perfectly regular, and some look frayed and worn. This is quite appropriate, since this is a rabbit's tongue, one of the most complex muscular structures in the mammalian body, and one in almost constant use.

Mammals' tongues have many uses, and come in many shapes and sizes. In humans they are used for speaking and, as in the higher apes, they also play an important part in gesticulation. Anteaters have long sticky tongues for eating ants and termites. Cows and giraffes have prehensile tongues for pulling at grass and leaves. And in virtually all mammals the tongue is used for manipulating food in the mouth and swallowing.

The microstructures on the surface of a tongue are varied according to the role of the tongue, or a part of it, and take the form of small bumps and lumps called papillae. Short broad papillae in the shape of squat mushrooms (fungiform), rings (circumvallate) or leaves (foliate) usually have taste buds. But the most numerous are the filiform (thread-shaped) papillae shown here, which give the tongue a more or less rough-textured surface. They are used to grip the food bolus as it is chewed, and to move it across the teeth and towards the back of the mouth as it is swallowed.

In rabbits, cats, cows and some other mammals, the filiform papillae are extremely abundant. They are arranged in regular arrays, like the teeth of a comb or brush. This is because the tongue is not only an organ for tasting and feeding, but also an organ for grooming. The long papillae help smooth and clean the hairs, keeping them parallel and unknotted to maintain the fur's waterproofing, weatherproofing and insulating properties.

Saliva brushed onto the hairs during tongue-grooming helps clean the fur, but it is also the source of typical cat- or rabbit-induced allergic reactions in some people, as it dries and flakes to become minute airborne particles.

This image so closely resembles a manufactured metal rasp that there is no mistaking the purpose to which the structure is put: it is for scraping and cutting. The crisp teeth are angled to their deepest gouge just on the curve of the fold, and the damage such a tool could inflict should not be underestimated. The mouth of every terrestrial snail or slug contains such a rasping device, called a radula. And even though each tooth is only 40 μm (0.04 mm) high, there may be up to 100,000 of them on a radula – and in these numbers they are devastating to susceptible plants.

The radula is a tongue-like organ just inside the mollusc's oral cavity. It consists of a belt of backwardly projecting teeth arranged in rows. The rasping action of the radula is superficially similar to the licking action of a tongue. Muscles slide the ribbon of teeth backwards and forwards over a block of underlying cartilage, which can be protruded out of the mouth and then retracted by its own muscles. The combined actions produce a continuous scraping of food particles into the mouth, from where they pass along the oesophagus into the stomach.

Although many gastropods (slugs and snails) feed on living plant tissue, the majority are detritivores, feeding on decaying organic matter such as fallen leaves, rotting timber, carrion and animal dung. A few are predatory on other molluscs.

The teeth of the radula are made of chitin, a tough carbohydrate-derived polysaccharide that also occurs in the hard exoskeletons of insects, crabs and other arthropods. They have a completely different chemical composition from mollusc shells, which are mostly composed of calcium carbonate. Despite their resilient strength, the teeth are constantly worn down, especially at the front of the radula, so rows of new teeth are generated at the posterior end and gradually move forward to replace them.

RIGHT
A rabbit's tongue is covered in minute projections called papillae. Some have taste buds, but the majority are for moving food about, and also for grooming.

LEFT
The soft fur of a rabbit is kept clean and smooth by regular grooming with the tongue's rough comb-like surface.

RIGHT
The curl of a slug's tongue, or radula, shows the many hundreds of sharp rasping teeth with which the slug scrapes away at its food.

LEFT
The soft and flexible body of a slug belies the mouthful of sharp strong teeth hidden beneath its slimy foot.

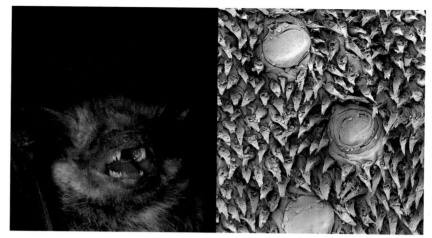

This architectural structure could be the high-tech roof of an airport terminal building, or the many-faceted bit of an oil drilling tool. Just as these structures reflect the function they have to carry out, so too does the business end of a fly's tongue.

Insect mouthparts come in two broad types: large jaws for biting and chewing, and tubes for sucking through. The sucking insects range from blood-feeding mosquitoes to nectar-sipping hoverflies. Pictured here is the broad bulbous tongue of a housefly, and its form is directly related to its function as a complex mop-like sucking device.

Houseflies scavenge on whatever they can find, from carrion and animal dung to fallen fruit and human food. They feed using a technique of external partial digestion, where the breakdown of the food into nutrient components is started outside the body rather than inside the gut. To achieve this, they start by ejecting the contents of their crop (upper gut) onto whatever they have landed on. This fluid contains complex enzymes that immediately start dissolving the food.

Each of the parallel-ribbed zipper-like bands covers a cylindrical tube. It is through these tubes that the crop contents are expelled, and it is up them that the resulting half-digested gloop is sucked. Resembling the suckered tentacles of some alien octopus, the divided tongue allows capillary action to work with the muscular sucking which will eventually draw up the insect's food into its body.

Unfortunately for humans, along with the digestive enzymes secreted onto the fly's intended food come the remains of the fly's previous meal. And when the fly lands on your Sunday roast as you leave it to stand for a few minutes, it might have just flown in through the window, having had an equally filling taster a short while ago on the dog dung on the pavement outside. Not a pretty thought.

The background texture could, at first sight, be a coir doormat or a bristly boot scraper, but the fried-egg motifs don't really seem to fit. In fact the likeness to a bristly mat is apt, because this textured surface has a similar use. Inside the back door, the stout bristle tufts are designed to grip mud and scrape it from shoes pulled against them. Inside the mouth of the pipistrelle bat, the bristly tongue helps grip insect prey, perhaps only half-caught in the animal's jaws.

The pointed projections, like those on the rabbit's tongue opposite, are filiform papillae, and they cover most of the bat's tongue. Sensitive to pressure, they give the tongue a sandpaper roughness to grip food and manoeuvre it down into the gullet. This prey-holding capability is very important in an animal that must keep its limbs outstretched whilst on the move. The fast-moving bat is trying to hunt fast-moving prey, and despite its well-developed echolocation sense, fluttering moths and flies are frequently caught just by the tips of their wings. The bat's nearly prehensile tongue will prevent many squirming insects from escaping.

The broad domed leathery-looking mounds are taste buds, called fungiform papillae because of their vague mushroom shape under the dissecting microscope. In bats (and other insect-eating animals), taste is not so much for gustatory pleasure as for staying alive. Bats do not taste their food to find a favourite flavour or distinguish between subtle scents – they use it to avoid being poisoned. One of the best insect defences against being eaten is to taste foul, by sequestering (harvesting and storing) potent and poisonous chemicals from their food-plants. The bat's decision to swallow or spit out is not taken lightly.

RIGHT
The tip of a fly's tongue comprises a complex series of convoluted sieves and corrugated channels through which its part-digested liquid food is sucked up.

LEFT
On landing on its food a blowfly extends its articulated tongue and probes with the soft spongy tip.

RIGHT
The rough texture of a pipistrelle bat's tongue helps it grip struggling insect prey, while the broad taste buds decide whether or not it is good to eat.

LEFT
Sharp teeth are also used to grip the bat's prey, mostly moths and beetles, caught in flight.

OPPOSITE
The convoluted skin folds of a spider's abdomen.
See page 60

OPPOSITE
Taste buds and rough-gripping surface of a bat's tongue.
See page 51

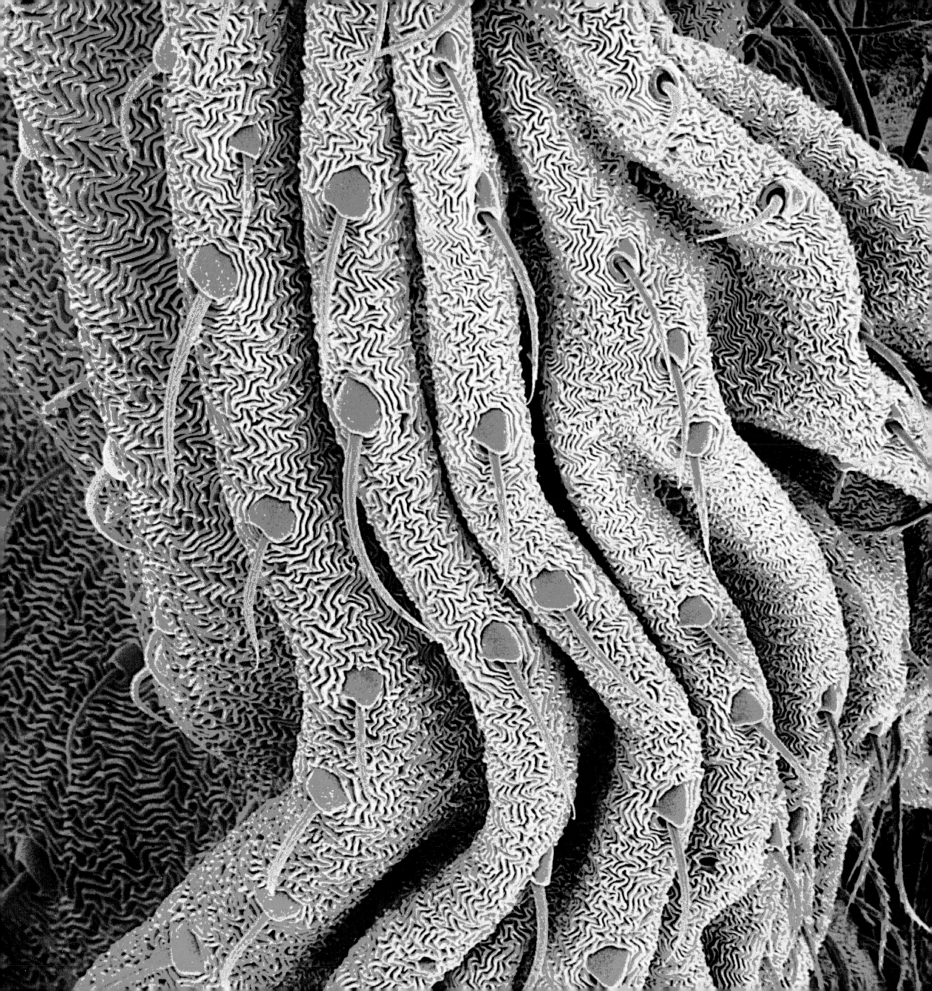

OPPOSITE
The net-like feeding and breathing mesh of a sea squirt's gills.
See page 60

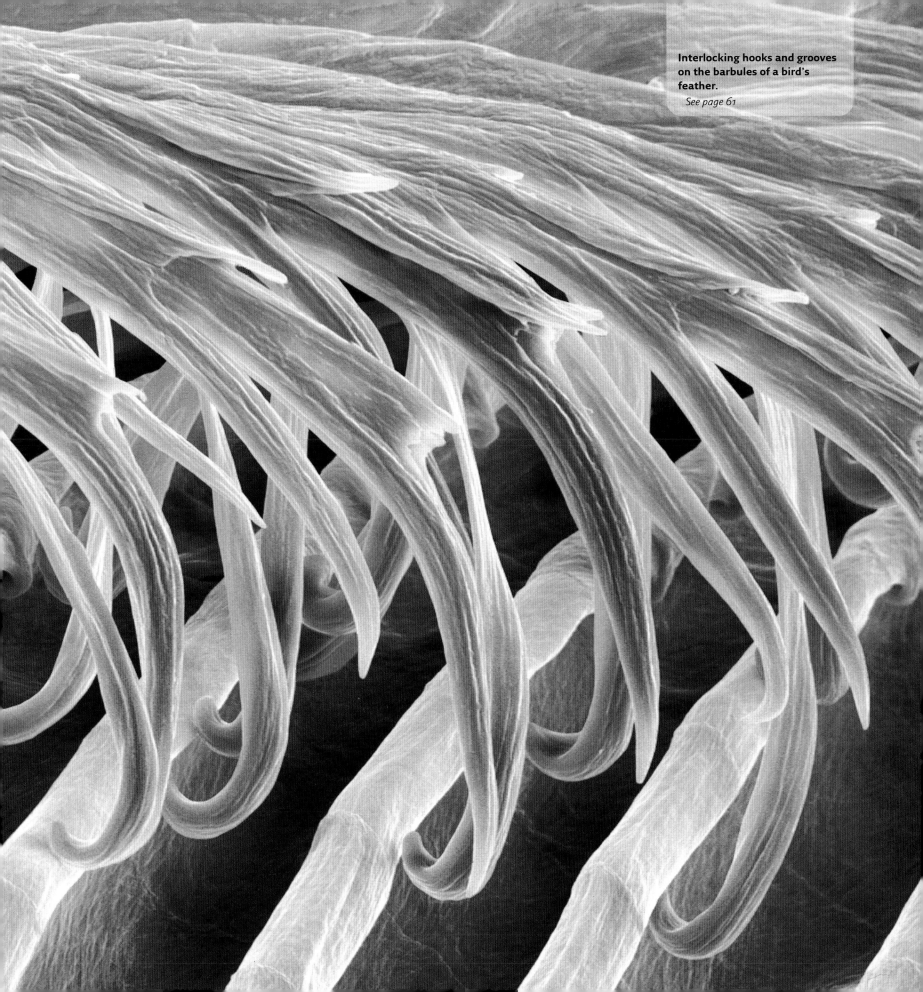

Interlocking hooks and grooves on the barbules of a bird's feather.
See page 61

OPPOSITE
The raised lacy covering of a foxglove seed.
See page 61

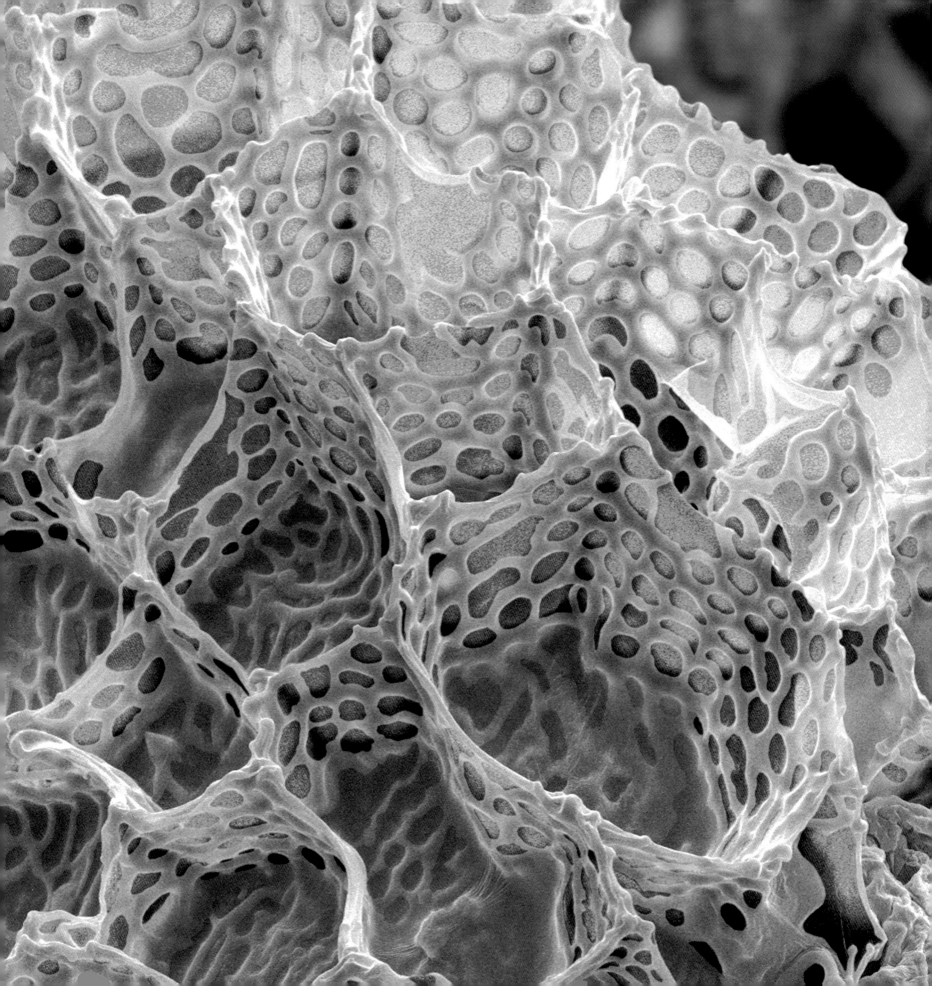

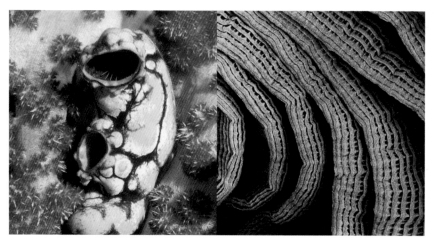

Wrinkled folds of this material resemble towelling, but the presence of many hooked thorns would make drying with it an unpleasant prospect. This is not, however, a bathroom fabric. It is the folded cuticle on the body of a spider.

Spider skin, like that of insects, is made up primarily of the polysaccharide carbohydrate called chitin, bound together with proteins in layers to create a tough laminate structure. In insects, hard plates of thick cuticle are laid down like pieces of armour, hinged by thinner, more membranous joints. In spiders, there are also tough plates on the cephalothorax (the front segment of the spider, housing the eyes, mouthparts and limb sockets) and cylindrical tubular plates along the legs, but the abdomen is much softer, more flexible, and prone to baggy folds.

In this respect, spiders are similar to other insect larvae, which tend to be soft and fleshy. In caterpillars, soft skin is important to cope with an ever-increasing bulk during this eating and growing portion of the insect's life cycle. Nevertheless there comes a point in the larval development when the skin has to be moulted away to make room for a new larger one. Spiders do not have a nymphal stage, but they, too, grow in spurts with each meal, and since their victims tend to be relatively large they can accommodate sudden large inputs of food by maintaining a softer and more flexible abdomen.

This does, however, mean that arachnologists have to store spider specimens in alcohol to preserve the soft bodies, a process that is more fiddly than the simple pinning of dried insect specimens enjoyed by entomologists.

These folds of material look like a raffia beach mat gathered into rough pleats, or a linen sheet just tumbled from the cupboard. The warp and weft of the coarse weave are clearly visible. But this structure is not a woven fabric; it is the internal gill mechanism of a marine organism called a sea squirt.

Ironically, this group of creatures gets its scientific name from the fact that they have the appearance of being wrapped in cloth tunics – they are tunicates. Doubly ironic is the fact that tunicates are amongst the only animals capable of manufacturing a substance similar to cellulose, the mainly plant product from which paper, cotton and linen are made. This is laid down in an outer sheath around the tunicates' bodies.

Adult tunicates have a simple tubular or bag-like form, with two openings for the passage of water. Water sucked in through one siphon is filtered through the netted folds of the gills, which are thus mainly for feeding rather than

breathing. Minute plankton are trapped in mucus and wafted by tiny hairs (cilia) down into the gut. The water is then ejected through the other siphon, called the atrium.

Adult sea squirts are soft blob-like gelatinous animals, permanently attached to coral or rocks on the sea bed. The larva, on the other hand, is free-swimming. It looks very like a tadpole, and offers a tantalising perspective on the evolution of larger animals. Microscopic examination shows that it has a stiff rod running through its tail, called a notochord. The notochord is lost when the larva lands, head-first, on a surface and develops into the adult. In other animals the notochord of the larva or embryo is not lost, but instead develops into the spinal cord and backbone. Tunicates, it turns out, are the best clue we have to a common ancestor from which all vertebrates – fish, amphibians, reptiles, birds and mammals (including humans) – have evolved.

RIGHT
Soft flexible folds in the integument allow the spider's body to increase suddenly in size after each meal. The body of the spider is also covered with tactile sensory hairs.

LEFT
A pale camouflaged crab spider waits in a flower for its insect prey, often much larger than the spider.

RIGHT
The intricate folded net of a sea squirt's gills sifts microscopic plankton from the ocean water pumped through its body.

LEFT
Sea squirts have a simple bag-like body with two openings, one sucking in water, the other passing it out.

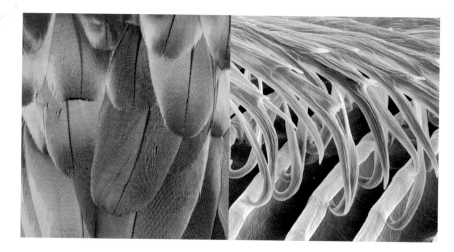

Rake-like claw hooks curl to grip the pale ribs beneath. Their tips are not sharp, instead clinging softly like Velcro. They are tiny hooked strands that make up a feather, and their combination of soft touch and firm grip gives the feather its shape and aerodynamic ability.

Among living creatures today, feathers are found only on birds, although they have also been found on some dinosaur fossils. Feathers give birds their beautiful plumage: they provide a huge variety of colours; they insulate against heat and cold; they protect against rainwater and provide buoyancy for swimming; they provide quills and plumes for courtship and territorial display; but primarily they provide birds with their most distinctive characteristic – flight.

Feathers are remarkably light, but very strong. Part of their strength comes from their composition of tough pleated protein fibres called beta-keratins, which form resilient structures much tougher than mammalian hair or fur, which is made of non-pleated alpha-keratins. However, the true strength of a feather lies in its microstructure.

Feathers are evolved from simple hair-like filaments, which have become branched and sub-branched to form flat blades. A feather comprises a central spar, called a shaft or rachis, off which are parallel branches called barbs. These are further branched into barbules, and the flat structural integrity of the feather blade arises from the mechanism by which these barbules grip each other. The parallel barbules on one side of a barb possess a series of small hooks, while those on the other side are grooved. All along the vane of the feather, hooks from one barbule grip the grooves on the next barbule.

Although each hook exerts only a light pressure, the combined grip of the aligned barbules holds the feather together. This lightness of grip also enables a bird to groom its feathers, removing wrinkles and rejoining splits and splays.

Looking like the starched lacy ruff of an Elizabethan high-fashion collar, these folds are not stitched fabric but a structural evolution for catching the wind, part of the husk of an airborne foxglove seed.

A plant seed is composed of three parts: the embryo, the endosperm (food store) and the seed coat or testa. Very often the tough seed coat is intricately patterned, and these distinctive patterns enable botanists to identify plant species in drilled cores of soil, peat or ice, in fossil deposits and in forensic samples.

The differences in seed pattern reflect the different strategies that plants have evolved for spreading their seeds around the globe. Animals and birds carry off hooked seeds (burs) in their fur or feathers. They also eat fruits, later passing the seeds in their faeces or spitting them out. Tough seeds like nuts and acorns are often deliberately buried as food stores and germinate when forgotten. Many, however, are simply dropped by the plant, and are moved by water, by soil disturbance, or in the air.

The lightest and smallest seeds, such as those from some orchids, are mere dust, blown about by the wind as aerial plankton. Others have mechanical attachments to catch the wind: sycamore have wing blades, while dandelion have feathery parachutes.

Foxglove seeds are intermediate: they are just small enough to be blown about by the wind, and are delicately sculpted with lacy ridges to help catch the air currents. On landing, immediate germination is not guaranteed, and most seeds can remain dormant until circumstances improve. Warmth from the sun or moisture from rainwater may indicate the correct growing conditions, a cold snap may show that winter really has arrived, so preventing a false alarm caused by a warm autumn, or the digestive juices in the belly of bird or beast may signify the passage of time and the corresponding passing of distance between the offspring seed and its parent.

RIGHT
The barbules of the feather vane are made up of hooks and grooves that hold them in place, producing a tough but flexible blade.

LEFT
The overlapping forms of feathers make them light but strong. As well as diverse colours and patterns, feathers provide protection against heat, cold and wet.

RIGHT
The convoluted casing of a foxglove seed helps it catch the air and disperse on the wind.

LEFT
The foxglove is a biennial plant, taking two years to flower: the seed produces nutrient-forming leaves the first year, then a spike of the distinctive flowers the second.

OPPOSITE
Airborne spores of the rose rust fungus.
See page 70

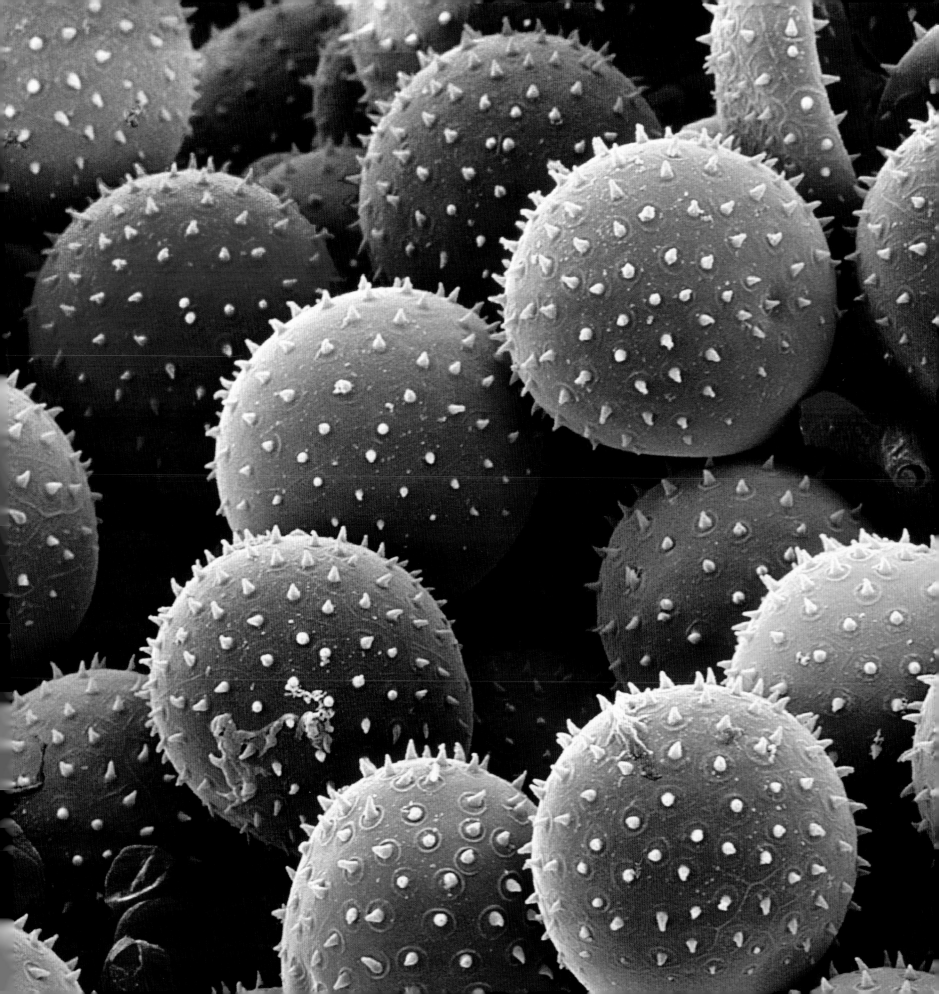

The delicate sculptured shape of these pollen grains helps them stick to the bodies of insects.

See page 70

OPPOSITE
A pollen grain embedded in the stigma of a flower.
See page 71

OPPOSITE
Pollen tubes germinating from grains of lily pollen.

See page 71

This pick-and-mix assortment of marshmallow balls may look good enough to eat, but it is they that will do the eating. These tiny fungal spores, each only a speck of powder, will start devouring a plant from the inside when they land on the leaves.

Unlike green plants, which invest in relatively low numbers of large seeds, each one well stocked with food stores for the developing embryo and protected by a tough outer casing, fungi have evolved a different dispersal strategy: vast numbers of flimsy spores, each containing little more than the genetic material. The trade-off between size and number works for the fungus because some of the multitude spores are almost certain to find a suitable place to grow.

Fungi grow on dead and decaying organic matter, including fallen leaves, timber, carrion and animal droppings, but large numbers also attack living plants. Like other diseases, they may attack just a single host. Rose rust fungi attack rose leaves, causing the rust-brown blotches that give them their name.

When a spore lands on a leaf it grows a small tubule called a hypha, which must penetrate the surface to gain access to the plant's cells in which the fungus will feed. The hypha 'feels' its way until it finds a stoma, a leaf pore opening, through which it can insert itself. Once inside the leaf the hyphae branch and spread, digesting the plant from the inside and absorbing the nutrients. This may eventually kill the plant, by seriously depleting its cell contents or by blocking the plant's stem tubes, through which water and nutrients flow between root and leaf.

Eventually the growing fungus reaches the point where it is ready to produce its own spores. Instead of creating a large fruiting body like a toadstool, rusts sprout small spore-bearing fronds, called conidiophores, out onto the leaf surface, giving the infected plant a dusty appearance as the spores (or conidia) are blown away.

These geometrically spiked globes, each bearing over 350 pointed projections, look like impossible mathematical models or complex three-dimensional cogs in some bizarre machine. They are but three grains of pollen.

Pollen is the plant world's equivalent of sperm: the carrier of the male genetic material that will fertilise the female plant's egg. Each grain is only a fraction of a millimetre across; the smallest is thought to be forget-me-not, at 6 μm (0.006 mm) in diameter.

The tough outer protective coat, called the exine, is composed of a very stable and decay-resistant substance called sporopollenin, and it is so beautifully and intricately sculpted that many plant species can be identified from just a single grain, a particularly helpful feature when they are found preserved in fossil remains and in soil, peat and ice samples.

Although a few plant species are self-pollinating, most plants benefit from ensuring cross-pollination, where pollen from the male part of one flower is only transferred to the female part of another.

Pollen makes this journey by one of three routes. Many trees and grasses produce pollen grains so small that countless billions of specks are wafted into the air at random. The sheer volume of pollen production virtually guarantees the final arrival of some on every female flower. This is the airborne pollen of hay-fever allergies: streaming noses and eyes, wheezing, coughs and sneezes. A few aquatic plants release pollen directly into the water in a similar manner.

The third mechanism is more convoluted, and it is the reason that beautiful and elaborate flowers have evolved – to attract pollinating animals who will unwittingly transfer pollen from one flower to another. This pollen is heavier and stickier and needs to be picked up and conveyed with accuracy and care. Insects are the usual carriers, but birds, bats and other animals are all involved, somewhere in the world.

RIGHT
The spores of the fungus consist of microscopic soft bags containing little more than the organism's genes.

LEFT
The rose rust fungus attacks rose foliage, turning the surface a brown rust colour.

RIGHT
The delicate sculpturing of pollen grains helps them stick to the bodies of pollinating insects.

LEFT
Bumblebees are supreme pollinators. And although they harvest much pollen to feed to their brood there is always enough brushed around on their bodies to pollinate the next flower they visit.

A small morsel seems lodged in the gently waving fronds of a sea anemone. It is poised on the edge of capture – as if the anemone were deciding whether or not to accept it. This is an important decision, for the morsel is a pollen grain, the male germ, and it has arrived on the female flower with its ovules awaiting fertilization within.

To complete a plant's sexual cycle, pollen must be precisely transferred from the male part of a flower, the androecium or anther, to the female part, the gynoecium or pistil. At the base of the pistil are one or more ovules, the plant equivalent of eggs, in one or more ovaries. Above this is a tall stalk, the style, topped with the pollen-receiving tip, the stigma.

Pollen is precious, both to the plant producing it and to the plant receiving it. This is especially true for plants pollinated by insects (and other animals), where a free-roaming intermediary is responsible for moving the pollen about.

All precautions are taken to ensure correct collection, delivery and receipt of the male genetic package. The evolutionary investment in producing attractive flowers to lure insects to them is matched by a similar investment in the engineering and design of the flower once the insect has arrived.

To ensure that pollen stays on the female flower once it has arrived, stigmas have evolved various shapes and forms. Some are sticky with moist exudates to glue the pollen in place until fertilization is complete. Others, like this periwinkle, are covered all over with finger-like growths, called papillae, into which the pollen grains become wedged.

Now that the pollen grain has arrived, the plant can ill afford to have it knocked off by a later insect visitor. Fertilization of the female plant's ovules requires the pollen to remain on the stigma long enough to sprout a root-like tube down into the ovaries, for the final transfer of genetic information to take place.

A jumble of exotic mushrooms ready for the pot would look no stranger than this mix of twisted stalks and rough-skinned fruits. But it is not the stems that have given rise to the latticed heads, it is the now deflated and desiccated tops that have sprouted roots. Each lobe is smaller than a pin-prick: it is a pollen granule, and the root-like tube heading downwards is the tube through which it will deliver its DNA cargo to the female's ovaries for fertilization.

Once pollen lands on the correct female flower, it must transfer the male genetic material it carries, the DNA, into the egg cell. The uniting of male and female DNA produces the full genetic complement that enables the seed to develop and, if conditions are suitable, the new plant to grow.

When the pollen lands on the stigma a process of chemical recognition takes place, ensuring that only pollen from the same species of plant completes fertilization. If the pollen is correctly recognized, it begins to grow a root-like tube

down into the pistil, the female part of the flower. The pollen tube first penetrates the stigma, the knob-like pollen landing point at the tip of a flower. It grows down through the style, the stalk supporting the stigma, and eventually to the ovaries in the seat of the flower where the ovules (eggs) are waiting.

As the tube grows, it obtains nutrients from the female flower. Lilies have relatively small flowers and the tube may be only a few millimetres long. Agriculturally bred maize (sweet corn) is often quoted as having the longest pollen tubes, reaching up to 300 mm.

Inside the tip of the growing pollen tube, the male DNA is divided into two sperm cells. When the tip of the pollen tube reaches the ovaries, both sperm cells are released. One fertilises an egg, which will develop into the new plant embryo, and the other fuses with two female cells called polar nuclei, also in the embryo sac, to form the endosperm, the food store for the developing seed.

RIGHT
A pollen grain has succeeded in getting itself delivered to the correct sex of the correct plant species. The tendril-like growths on the female stigma hold the pollen grain in place while fertilization occurs.

LEFT
In return for moving pollen about, flowers offer insect visitors energy-containing nectar.

RIGHT
When they land on a female flower, each of these lily pollen grains sprouts a long tube to take the male genetic material down into the ovaries to fertilise an egg.

LEFT
The pollen tube must penetrate the long style, here surrounded by six stamens, to reach the ovules deep in the flower.

OPPOSITE
**Branched hairs and oil gland from
a lavender leaf.**
See page 80

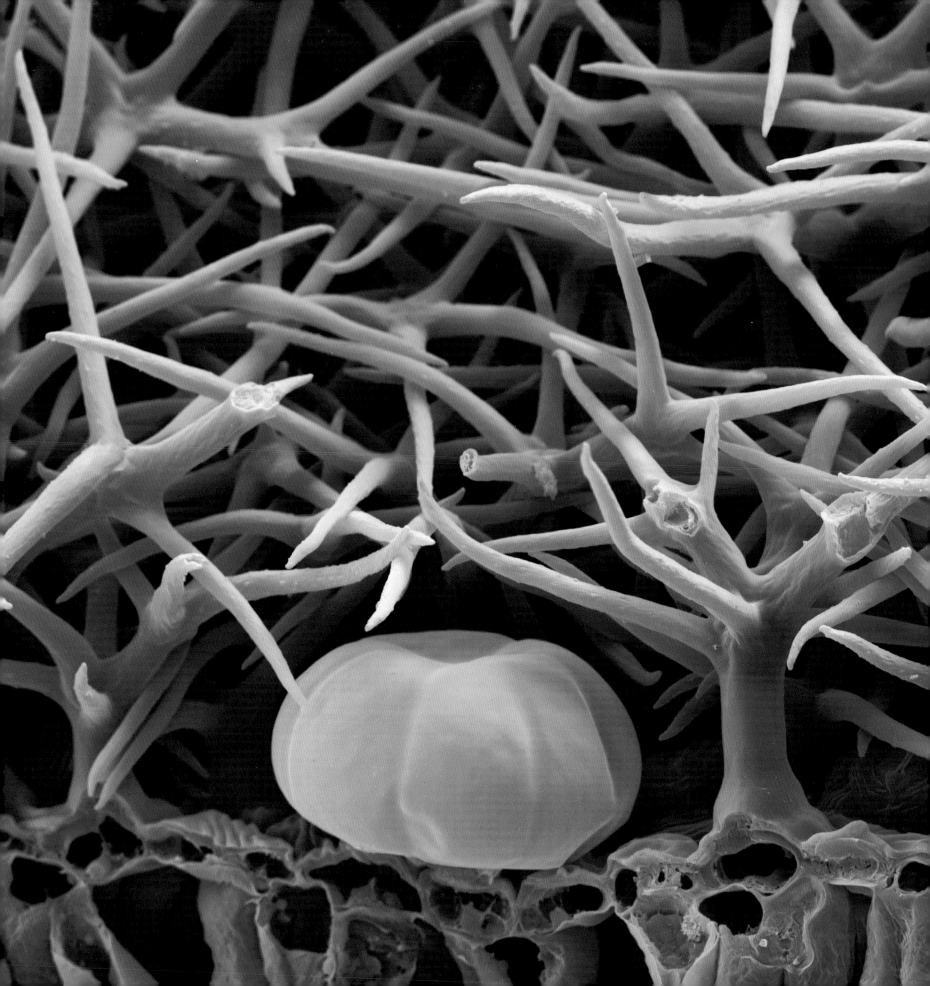

Flattened hairs covering the underside of an olive leaf.

See page 80

OPPOSITE
**Oil glands top the hairs covering
a clary sage leaf.**
See page 81

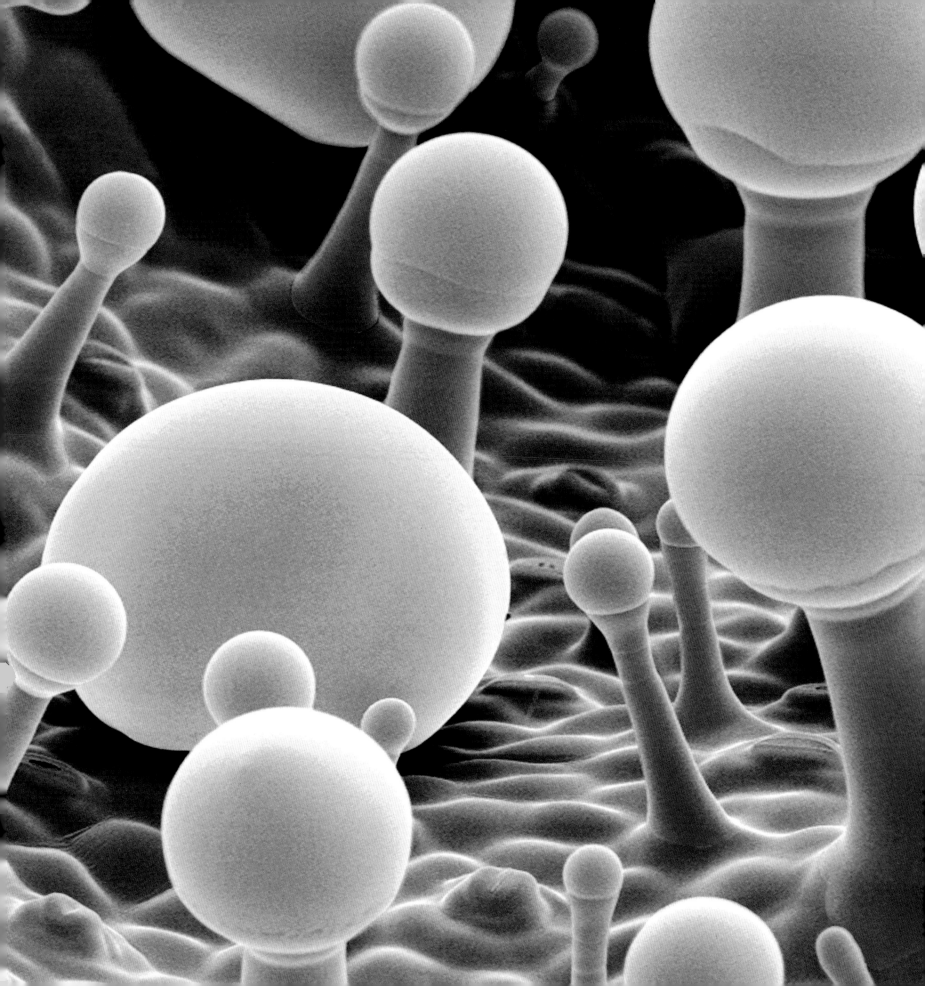

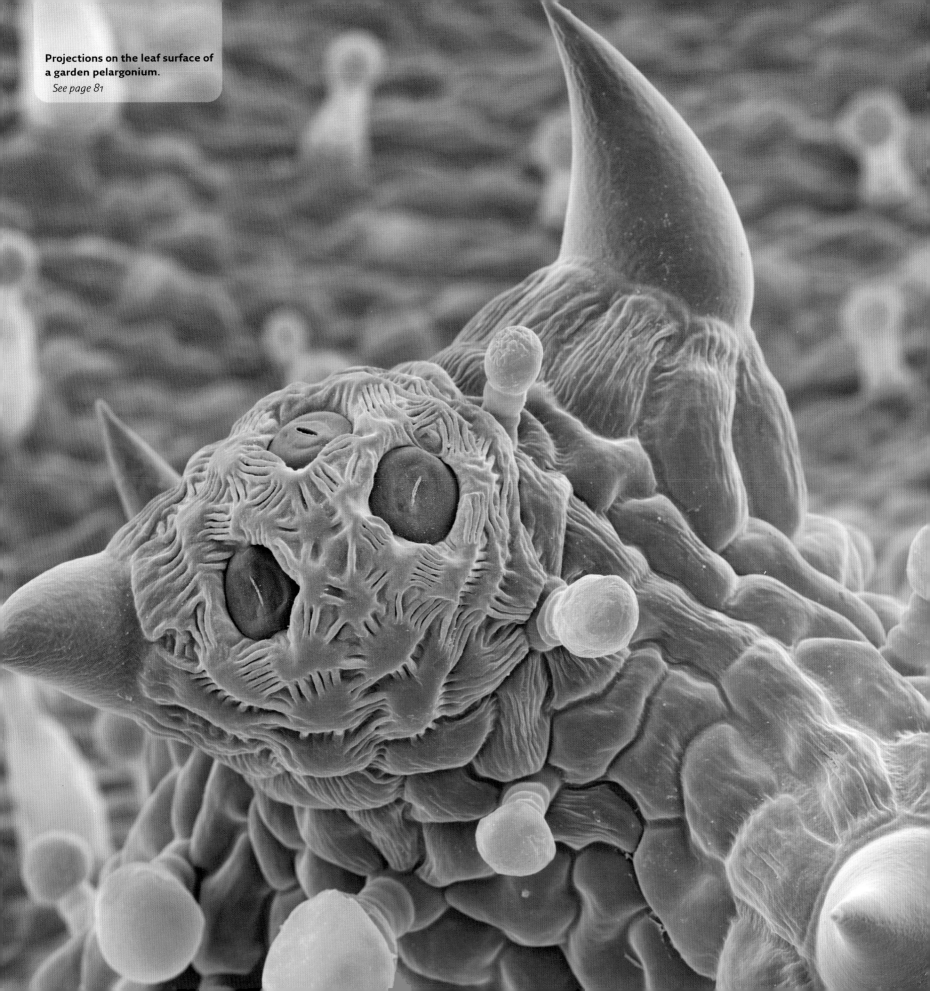

Projections on the leaf surface of a garden pelargonium.

See page 81

There is something eerie and ghost-like about this woodland landscape. The intertwined limbs seem too tight and the boles are too smooth. What looks like a forest of densely packed trees is in reality a blanket of branched hairs, called trichomes, that coat the surface of a lavender leaf. These hairs give the lavender leaf its soft downy appearance and have at least three distinct functions.

Hairs act as a physical deterrent to insect herbivores, which get entangled or cannot penetrate to the nutrient-containing cells of the leaf surface. Some larger herbivores find the hairs irritant or distasteful.

The blanket effect of the hairs keeps down water loss from the leaf, insulating the surface by trapping moisture in the complex branches and preventing it being blown away by the wind. Lavenders (and other downy-haired plants) typically grow in hot dry conditions where other plants are liable to wilt and die.

Pale hairs reflect sunlight, helping to keep the leaves cool and further preventing too much water loss through evaporation from the leaves.

The pale bag-like structure in the centre is a gland containing the potent oil that gives lavender its strong scent. The oil is a complex cocktail of small biologically active molecules, mostly terpenes and terpenoid alcohols.

Aromatic plants like lavender deter the attentions of herbivores with powerful-smelling chemicals. Leaf-eating insects have to cope with high concentrations of these substances, which may be poisonous unless they expend large amounts of energy metabolizing or excreting them. Aphids and other sap-suckers ordinarily have to drink relatively huge amounts of plant juices to get enough protein to grow, so even small amounts of these chemicals cause them problems. By concentrating the oil into a bag-like gland, lavender is also putting larger herbivores off, by pro-ducing tongue-tingling star-bursts of bitter flavour as the animal chews its way into the leaf body.

These delicate pale turquoise doilies on the under surface of an olive leaf are more than just a prettily coloured decoration for the tree. They have nothing to do with dining etiquette, they are neither woven nor crocheted. The strange flattened tentacled discs are specialised trichomes, and they are a perfectly honed adaptation to life in the harsh arid conditions of dry mountain slopes during a Mediterranean summer. These are scale-like leaf hairs that protect the underside of the leaf, and thus the whole plant, from desiccation.

Water loss is the life and death of a plant. Controlled water loss through evaporation from the leaves is one of the basic principles that allow a plant to live and grow. As water vapour is excreted through small pores, called stomata, in the underside of the leaf surface, this creates a negative pressure in the water-carrying vessels of the plant's stem, helping to suck up more water and mineral nutrients through the roots.

Plants are able to control some of this water loss by opening or closing the stomata. But since these pores also allow ingress of gases, particularly carbon dioxide, which plants absorb and metabolize during the process of photosynthesis, prevention of water loss and the creation of food are sometimes in competition with each other.

The olive tree can survive in seasonal drought conditions that would kill other plants. Well-drained soils and the drying effect of the wind cause softer plants to wilt beyond recovery. The olive tree's leaves, however, are coated beneath with a protective barrier of trichomes, effectively creating a blanket under which a moist layer of air is not blown away so fast by the wind. This allows the stomata to remain open for gas exchange, but keeps down water loss.

RIGHT
Branched hairs protect a lavender leaf from evaporative water loss, while a gland full of bitter-tasting lavender oil deters herbivores.

LEFT
Up above the herbage, however, the lavender flowers are exuding another scent to attract bee pollinators.

RIGHT
Scale-like trichomes protect the underside of an olive leaf from water loss through the stomata, several of which are visible in the gap just above the centre.

LEFT
Olives grow in hot dry places where wilting is a constant threat, and their silky leaves are an adaptation to the harsh environment.

This science-fiction landscape of opaque domed lightbulb trees would confuse the most hardened of astronauts. Looking soft and gelatinous, but at the same time delicate and fragile, each sprout looks like it might burst if touched. And burst it would, for this forest of blobs is part of the clary sage plant's defence against being eaten.

Each pale globe mounted on its short stalk (another form of trichome, like those shown opposite) is a structure evolved from simpler hairs on the plant surface. The spherical tip is a specialist gland that secretes an aromatic oil, distasteful to smaller nibblers such as insect caterpillars.

Clary sage is a native of Europe and Central Asia, and has long been noted for its aroma, variously described as smelling like sweat, spice or hay. The leaves have been used as a vegetable, in salads and as a tea drink, but it has been most widely cultivated for flavouring. The leaves are said to be effective against stomach problems, like wind and indigestion, and as a tonic to relieve period pains and premenstrual tension. Once used to flavour beers and wines, today it is harvested to distil an essential oil from its leaves. The oil is used as a carrier base in the perfume industry and as an ingredient in aromatherapy, where it is said to add a euphoric element to the scents.

The oil contains a mixture of different chemicals, including, in particular, sclareol, a complex compound very like cholesterol. Like many aromatic plant substances, it has been found to have widespread effects on the human body and is being intensively studied for its medicinal properties. Sclareol inhibits cell growth, and is being investigated for its ability to prevent cancers from growing. It also has some antibacterial properties.

Some strange creature, a chimaera, half beast, half plant, seems to be rearing up, threatening with its large claw horns. At high magnification, the apparently smooth surface of a pelargonium leaf is revealed in all its undulating contours of bumps, warts and thorns. Despite the plant's relatively soft and smooth appearance to the naked eye, pelargonium leaves have a distinctly tough and bristly-paper feel when crumpled in the hand. It is these horns that make them so.

This type of leaf projection is a defence against herbivore attack. The thorns are tough and irritating on the palate of vertebrate grazers, and indigestible or damaging to the mandibles of insect browsers such as caterpillars. The mushroom-shaped stalks are gland hairs that produce the aromatic scent for which pelargoniums are cultivated.

Plants and herbivores (particularly insects) wage a constant evolutionary war of attrition against each other. The synthesis and concentration of strong-smelling chemicals by plants is a defence against attack. Grazing animals have to expend large amounts of energy detoxifying what they eat.

Pelargoniums (sometimes previously called geraniums) are well known as garden plants. Most are native to southern Africa, but a few occur in the Middle East, Madagascar, Australia and New Zealand. Like many exotic plants now used in gardens well away from their native ranges, pelargoniums are not much troubled by insect pests because their chemical defences successfully deter generalist browsers, which find easier and tastier food elsewhere. However, the geranium bronze butterfly has caterpillars that not only survive the chemical warfare but have come to feed solely on these plants. Although native to South Africa, they have been accidentally transported around the world by the horticultural trade in pelargonium plants, and they now threaten to become a major pest of these popular garden plants.

RIGHT
The pale mushrooms that cover the surface of the leaf are oil glands, which secrete a mixture of strongly aromatic chemical compounds.

LEFT
The flowers of clary sage contain high concentrations of essential oils and are harvested for flavouring and therapeutic concoctions.

RIGHT
Looking like some mythical horned creature, this fleshy blob is a raised pustule on the leaf of a garden pelargonium.

LEFT
Although appearing soft and downy, pelargonium leaves are tough and fibrous to the touch.

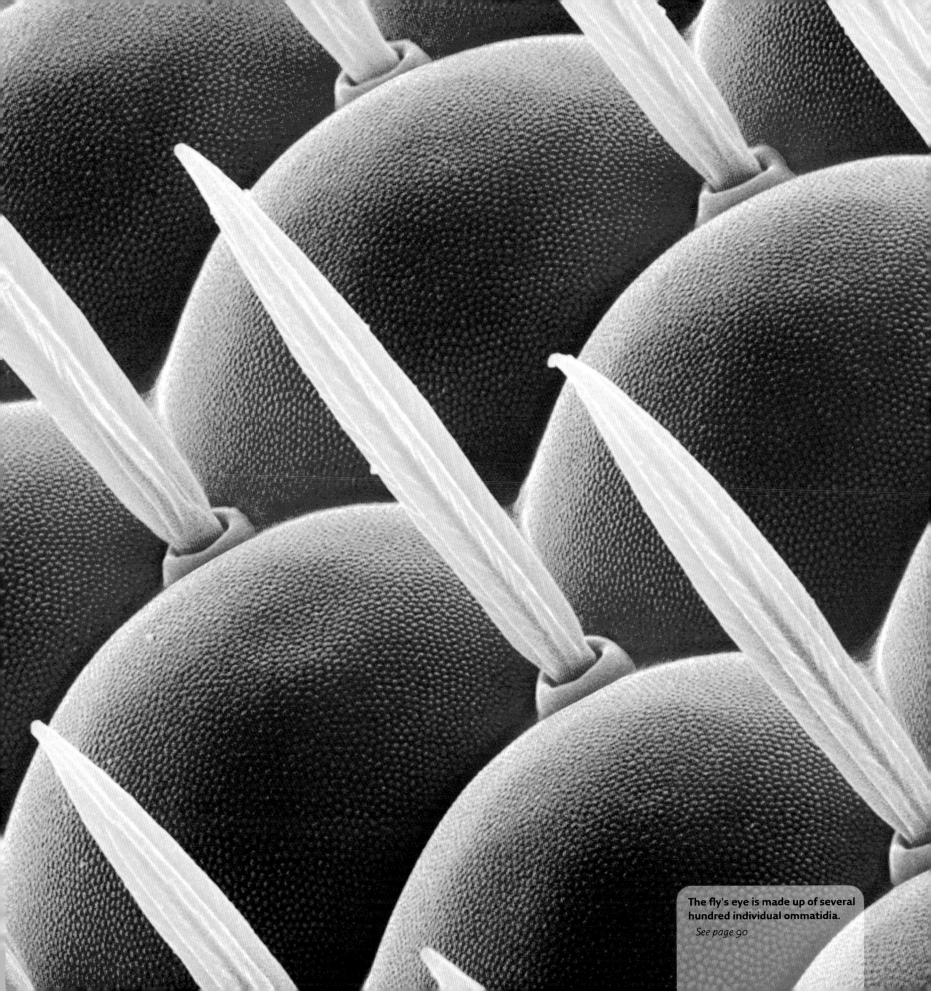

The fly's eye is made up of several hundred individual ommatidia.
See page 90

OPPOSITE
Multi-domed surface of a brittle star body.
See page 90

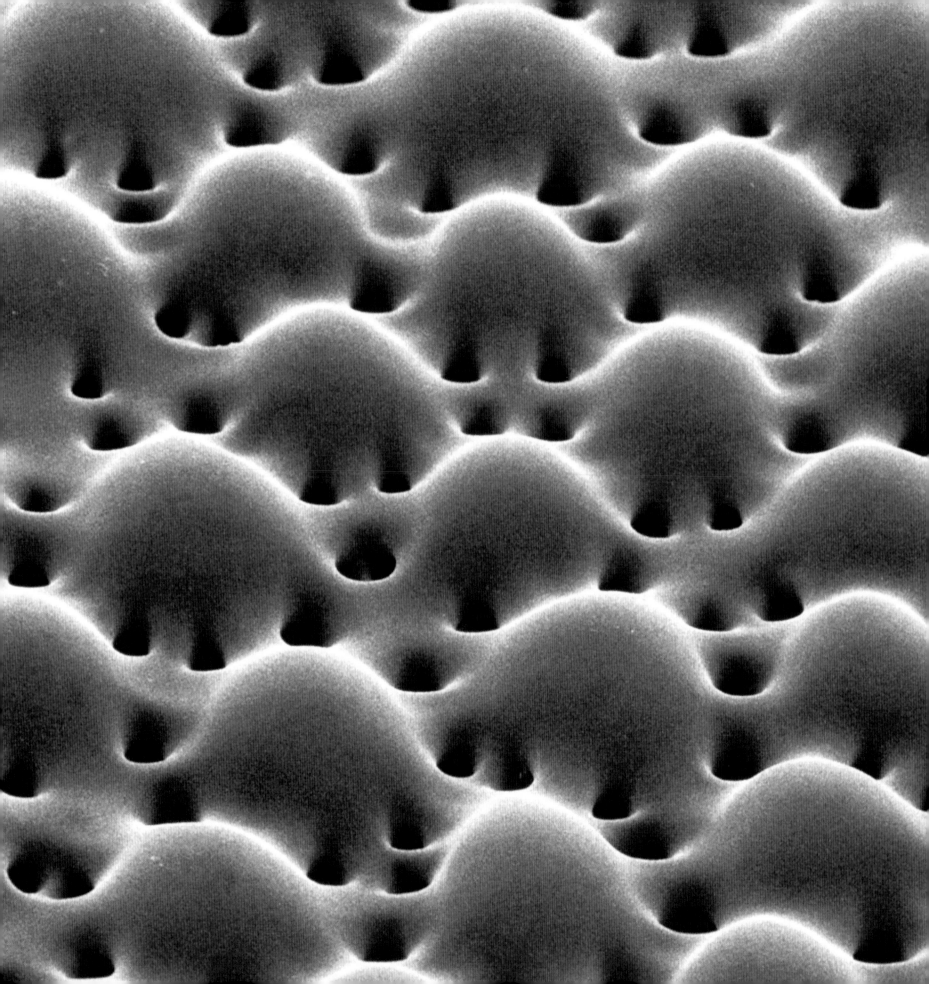

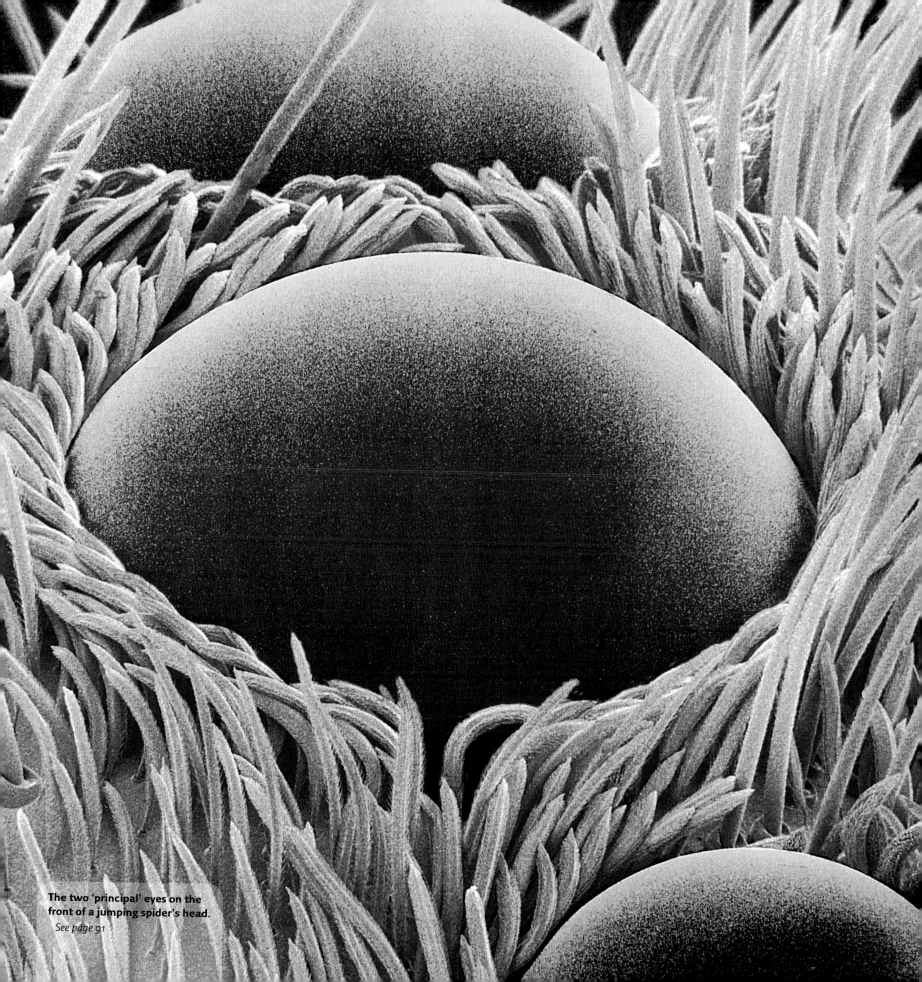

The two 'principal' eyes on the front of a jumping spider's head.

See page 91

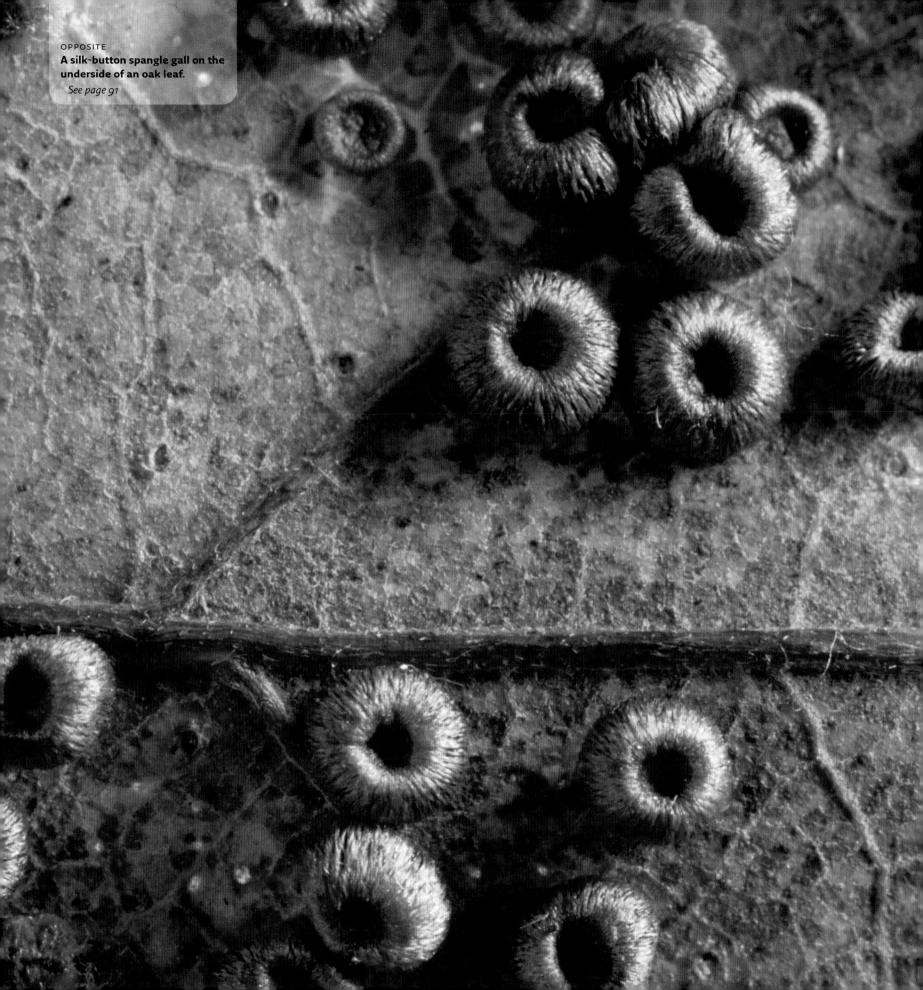

OPPOSITE
A silk-button spangle gall on the underside of an oak leaf.
See page 91

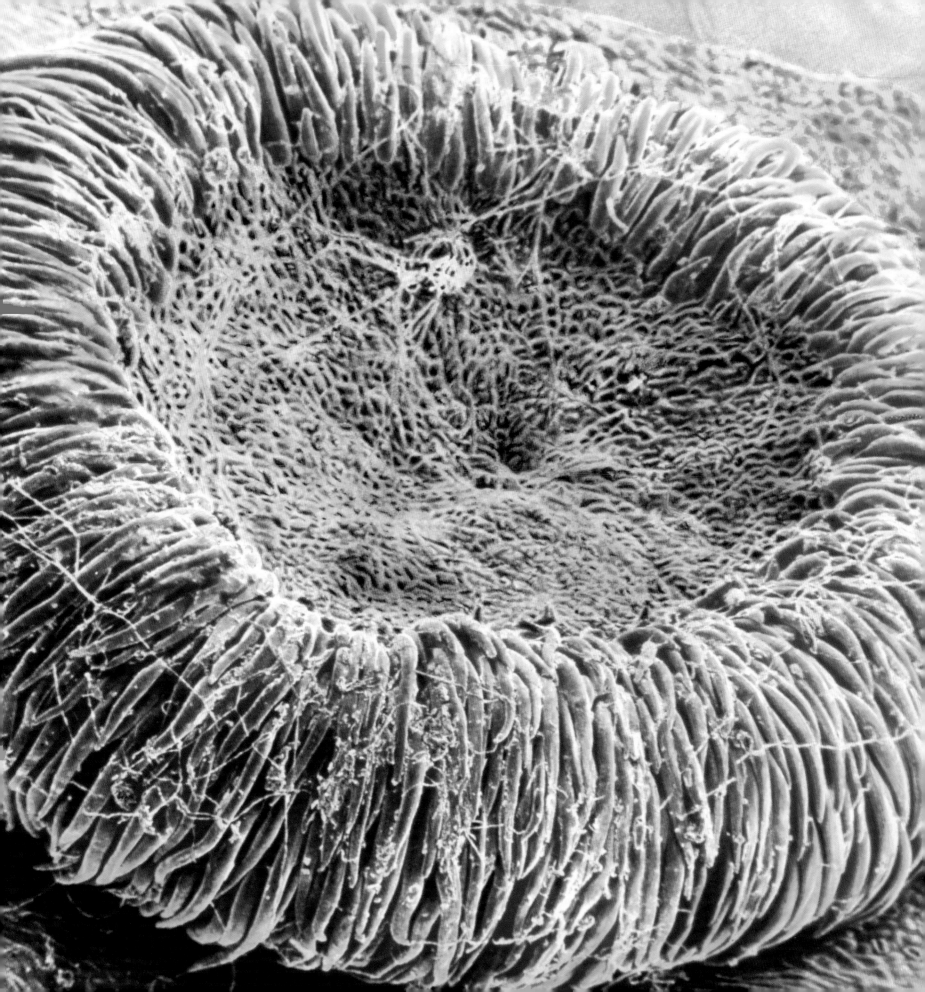

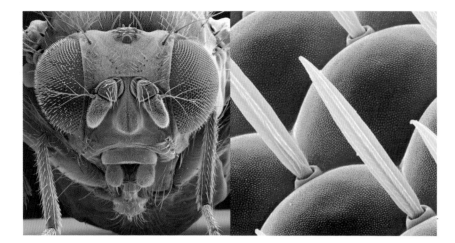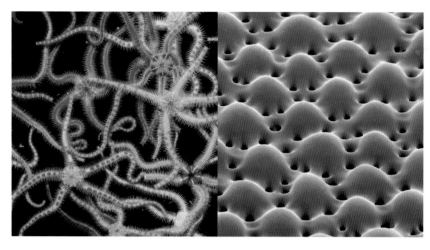

Insect compound eyes are a marvel of minute multiple construction. Each consists of many individual units called ommatidia, stacked together in a dome formation. An ommatidium is made up of a transparent cuticle to protect it, a curved lens to focus the light, a thin column of light-sensitive cells called a rhabdome and finally an optic nerve fibre to take sensory signals to the insect's brain.

Each ommatidium functions as a separate unit. The image produced is very vague, nowhere near as sharp as the crisp pinpoint detail created by avian or mammalian eyes. It is by the overlapping of the multiple images of so many individual lenses that the insect can produce a usable picture of the world around it. But even this final image is rather poor. It is more a mosaic impressionist interpretation of light and dark patterns, rather than a full 'picture' like a photograph.

While its low-resolution sight does not give the fly much in the way of an artistic interpretation of the world, this type of visual mechanism is extremely sensitive to movement. Any object that moves, even the slightest bit, needs just to affect the light entering one single ommatidium for a warning nerve signal to be sparked off.

Outwardly, the eye is covered with a more or less regular hexagonal array, showing how the ommatidia are fitted together. This fruit fly has about 250 in each eye, a relatively low number, but nevertheless allowing the insect to discern light and dark areas enough for it to find fallen fruit beneath trees. Dragonflies, by contrast, are supreme aerial hunters and need visual acuity at its peak – so it is no surprise that the largest of these have upwards of 30,000 ommatidia in each eye.

The combination of undulating surface and tiny holes suggests this might be some drainage system. Or perhaps it's a sieve. Or maybe a porous non-stick pizza pan. These minute domes occur across the bony body of a brittle star, and their precise use kept scientists guessing for years. It was only when a researcher polished a few and tested their refraction properties under direct light that they were recognized as tiny lenses.

Brittle stars, thin-limbed relatives of starfish, are mainly nocturnal creatures, emerging from shelter at night to feed on plankton and other small prey. They get their name from their long snake-like arms, which look very fragile, connected to a relatively small disc-like body. Indeed, the ends of the arms are easily lost to predators, but can be regenerated.

Unlike starfish, which are rather ponderous movers, brittle stars use their undulating limbs to move quickly, especially to escape potential predators.

It has long been known that they are able to detect shadows falling over them, because they move away quickly whenever this occurs. What confounded scientists was the fact that these marine organisms have no eyes, or any other obvious visual apparatus.

Scanning electron micrography revealed that the surface of the brittle star is covered with lens-shaped domes, and it was postulated that this may give the animal some primitive sensitivity to light. The domes, like the rest of the tough shell of the brittle star, are made of a form of calcium carbonate called calcite. By testing the optical qualities of these calcite domes, scientists soon discovered that they do function as lenses, focusing light at the point below the brittle star's surface where nerve receptors lie. This means that the whole body of the star is in effect a compound eye, and it probably has much greater visual acuity than many other supposedly more advanced animals.

RIGHT
Despite their large number, the individual convex facets (ommatidia) only give the fly a rather coarse blurred vision.

LEFT
The hemispherical multifaceted compound eyes of a fruit fly dominate its round head.

RIGHT
Between 10,000 and 20,000 tiny domes cover the body of the brittle star *Ophiocoma wendtii*. Each is a small lens, focusing light on nerve cells beneath the shell, making the entire organism one huge eye.

LEFT
Brittle stars are related to starfish, but are more agile and fast-moving.

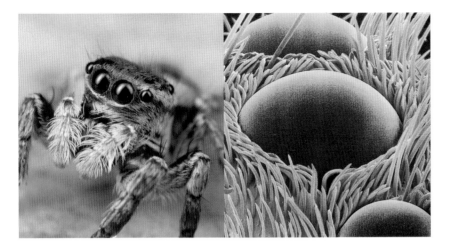

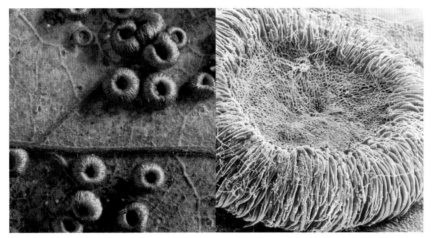

This row of orange balls, half-buried in long grass, is all the more striking because of the unblemished sheen of the orbs. Unlikely to have arrived in a line by accident, they look as though they are placed with aesthetic accuracy and artistic purpose.

In truth, their position serves a fundamental purpose for the jumping spider that possesses them, because these are two of the spider's four neatly aligned front eyes, and they are the key to its hunting success.

Jumping spiders no longer make silk web snares to trap their prey. Instead, they hunt by sight, pouncing on their victims in a sudden leap, often several times their own body length. For such an athletic feat they must accurately calculate their trajectory, which requires a precise measurement of the prey's size and distance. To do this, jumping spiders have acquired a visual acuity second to none.

Spiders have eight (or rarely six) eyes, usually little more than small light-sensitive domes that detect a shadow falling across them. But in jumping spiders the front four eyes are large, the middle two extremely so, often 100 times larger than in other spiders of similar size. This gives them accurate binocular vision, allowing them to carefully triangulate the position (and thus the distance) of an object in front of them. They have an internal lens, as well as the outer lens dome seen here, which transforms the eye into a telephoto device, bringing objects from about 2 cm to infinity into focus. They also have muscular control over the light-sensitive retina at the back of the eye, allowing the spider to change the direction of its gaze without moving its body. And within the retina there are four layers of rhabdomeres (light-sensitive cells), arranged to receive images from different parts of the light spectrum and giving jumping spiders excellent colour vision.

This strange bagel-shaped construction could be a cushion or a woven wicker garden sculpture. The curled edges look soft and spongy with a thousand arched tendrils, but what are they concealing beneath? Inside this elaborate construction is the grub of a tiny midge-like insect, a gall wasp called *Neuroterus numismalis*.

Gall wasps are related to wasps, bees and ants, but they have pursued a radically different lifestyle. When an egg is laid inside a plant, a cocktail of chemicals is injected too. The chemicals cause the leaf, stem or bud to start growing abnormally. When the grub has hatched it continues to produce substances which subvert the plant's normal growth patterns and cause a bulge called a gall to grow. Each species of gall wasp causes a different shape of gall, normally on only one host plant species.

This particular gall is called a silk-button spangle gall because of its close

resemblance to the fabric-covered buttons of showy dress clothes.

There is no doubt that gall-causers usurp the plant's energy and nutrients, but galls may also be the plant's attempt at damage limitation, containing the grub in one particular spot and preventing the wholesale tissue destruction caused by grazers like caterpillars. There may be scores of spangle galls on a single oak leaf and, potentially, millions on a tree, but this does not seriously impair the tree's growth or seed production.

Gall wasps have complex life cycles. In spring, only females emerge from the silk buttons, after they have fallen with the autumn leaves. They will lay unfertilized eggs on newly budding oak leaves, but this time the galls that develop are no more than indistinct swellings. These give rise to both males and females, which mate, and the resulting eggs produce the next generation in silk buttons.

RIGHT
A jumping spider has two large 'principal' eyes at the front of its head, giving it superb binocular vision to judge the size and distance of the prey.

LEFT
Jumping spiders leap to catch their prey. In case they fail they trail a line of silk and abseil to safety.

RIGHT
Within the protective swelling of the silk-button spangle gall, a tiny gall wasp grub is developing.

LEFT
The galls fall with the leaves in autumn, and in spring the emerging gall wasps will seek out new oak leaves.

Waxy mounds coat the petal of a primrose flower.
See page 100

OPPOSITE
**Clumps of waxy secretion coat
the body of a weevil.**

See page 100

OPPOSITE
Minute hairs coat the wing membrane of a mosquito.
See page 101

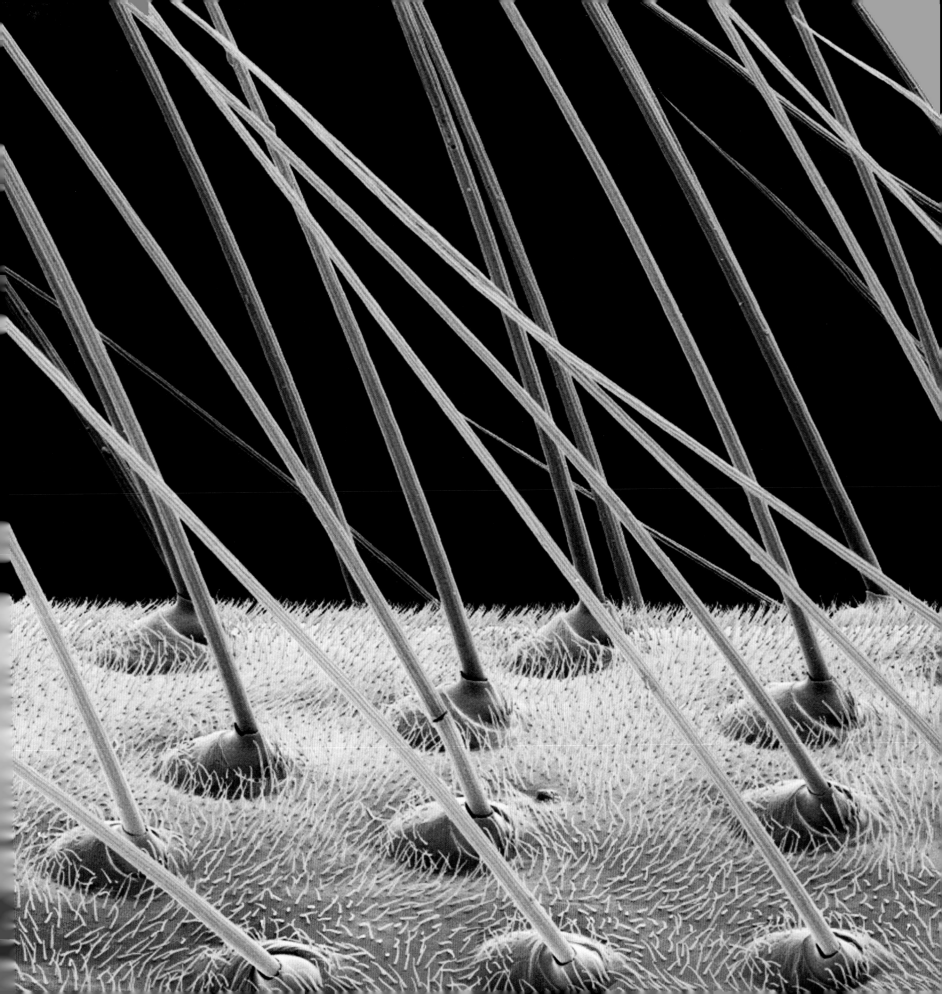

OPPOSITE
Insect-trapping hairs cover the leaf of a sundew plant.
See page 101

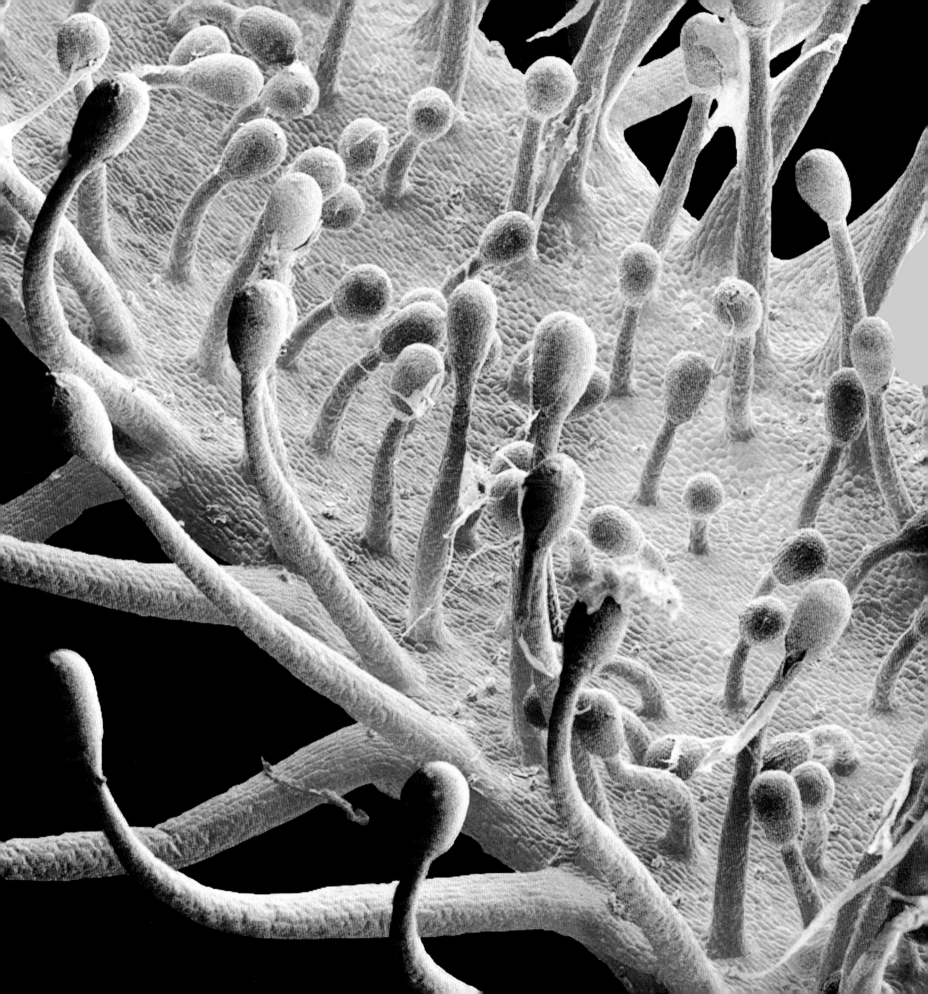

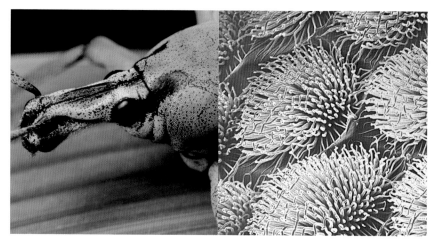

These gentle raised conical mounds look like dimpled latex foam, but what are they cushioning? They are certainly soft, but the only things they are likely to come into contact with are the feet of insects, because this is the surface of a delicate primrose petal.

Flowers have evolved in parallel with the pollinating insects that visit them, and have developed the large bright forms and strong colours we see today. Petals (and other parts of the inflorescence) are derived from highly modified leaves, most losing the green of the photosynthesis pigment chlorophyll in the process. Instead, they take their colour from other pigments stored in a series of waxy mounds called papillae.

The papillae are raised lobes, growing out from the epidermal cells of the petal. In different plants their shape varies from gentle domes, through the conical examples seen here, to long finger-like villi, and botanists are now starting to compare them when examining how plant species are related to each other and how flowering plants have evolved.

The waxiness has been attributed to the prevention of water loss, an important factor given that insect visitors are most active in the heat of the day, during which time the petals are most susceptible to wilt. However, it has been suggested that in some plants the papillae also secrete insect-attracting substances in lieu of nectar, or chemicals that modify insect behaviour to ensure that pollinators linger longer at a bloom.

The light-reflecting properties of the papillae may also be important, creating the soft sheen typical of petals, and contrasting with the higher shine of the surrounding leaves.

In nature, red, orange and yellow (through to white) flowers predominate, with blue and purple species in the minority. However, in plant groups that have received attention from gardeners, selective breeding has created a new series of different-coloured cultivars.

The delicate tuffets of tubular tendrils arranged in soft cushions could be the flowers of thistles or sea thrift, or the waving bodies of sea anemones packed into a rock pool. The flower resemblance is well known, for although these tufts are sprouting over the surface of a beetle, they are referred to as a 'bloom'.

Each clump is made up of a series of strands of a thick waxy substance extruded from the surface cuticle of the beetle. In life they take many colours, and like the scales on a butterfly's wing they can combine together to produce different patterns of stripes, blotches or freckles. Many beetles do have scales, but that covering has one disadvantage: if the scales are scraped off by a predator or (more likely) through wear and tear, they are not regenerated and the colour or pattern is lost. But if a wax bloom is brushed away, it is replaced within hours or days, and the beetle's camouflage or colours are returned.

In hot deserts, specialised beetles grow a white bloom on their backs, thought to help them keep cool by reflecting the sun's rays. White is not an easy colour to produce in pigment, and other white beetles are often so coloured because of reflective qualities in the cuticle.

The origins of waxy secretions may be functional, rather than decorative, because wax has one very important property – it is waterproof. On the whole, insects keep dry by being covered in water-repellent hairs, which at such small scale hold back the water by resisting its surface tension. A coating with clumps of hydrophobic wax also keeps water at bay, and this is used by many beetles that have partly aquatic habits – those that feed on pond-side vegetation or in the tidal zone of the seashore.

NANO FORM

RIGHT
The apparently smooth and even surface of a primrose petal is revealed as a series of conical bumps, called papillae, which arise from the epidermal cells.

LEFT
Primroses are simple open flowers, the petals forming a base for insects to rest on.

RIGHT
Clumps of waxy secretion coat a weevil shell and provide diverse colour patterns and built-in waterproofing.

LEFT
Weevils which live exposed, feeding on plants, are often strikingly coloured, either with mottled camouflage or to disguise the body outline.

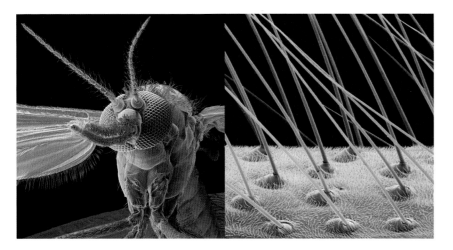
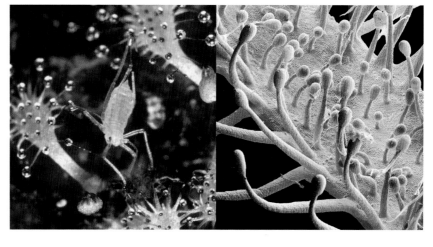

This outlandish landscape of tall green grass stems growing from a pink turf is about as far removed from the reality as is possible without the use of psychotropic drugs. Each stem is a tiny hair, called a microtrichium, and it is sprouting from the surface of a mosquito's wing.

Fly wings are usually perceived as being thin smooth transparent membranes stretched between radiating veins. This is a perfectly adequate simplification for comparing them to, say, the brightly coloured scale-covered wings of butterflies and moths, or the hardened leathery wing-cases of beetles and cockroaches. But under higher magnifications they are revealed as being covered with these microscopic bristles.

The precise function of microtrichia is still not fully understood. Different fly species have different arrangements, with longer or shorter spines, larger or smaller bare patches between the microtrichia, different densities and different patterns. The characteristic distribution of these hairs on the wings has proved a useful tool for entomologists when it comes to identifying groups of very similar fly species.

The longest-standing theory is that microtrichia aid insect flight. The fly wing is not just a simple aerofoil (like a kite) that translates the movement of air beneath its surface into downward thrust and thus gives it upwards lift. Instead, the flapping movement of the fly's wing produces downward thrust with each forward and backward movement of the wing membrane. This is achieved by the fly twisting its wings so that with each movement it catches the air to propel itself upwards and forwards. But between the finish of one movement and the beginning of the next, as the wing is flexed, there is a period when the wing stalls and no power is produced. Microtrichia, it is argued, help prevent stalling by creating drag – almost clinging on to flight by the friction of the microtrichia against the air.

What appear to be delicate marshmallow-tipped matchstick fingers growing from the long flat strand beneath are indeed a sweet sugary confection. But they are also deadly, for this is a sundew leaf, and these stalks are part of a highly efficient insect-killing mechanism.

In life, each lobed tip would be covered with a glistening drop of liquid, hence the dew of the plant's name. The droplets contain sweet sugars to attract insects, but they are also very sticky, and any visitor is quickly trapped. To ensure no escape, the sundew leaf then starts to move, curling around its victim (a process taking only an hour or so) and fully coating it in mucus. The insect is killed by asphyxiation as the thick liquid blocks its breathing tubes. This is followed by the much slower and more insidious process of digestion, as enzymes in the 'dew' start to dissolve the insect's body into a nutrient soup that the plant can absorb through special pores in its leaf surface.

This unusual behaviour allows sundews (and other insectivorous plants) to live in very nutrient-poor conditions, especially in the acidic waterlogged soils of bogs, fens and marshes. In nature, nitrogen, a key element of proteins, is a precious resource. Despite its abundance as a gas in the atmosphere, usable nitrates and nitrites must normally be absorbed by a plant's roots from mineral sources in the soil. Sundews get their nitrogen from the digested insect proteins.

Sundew flowers are simple, usually pale, affairs, on long tall stems high above the sticky leaves. This was always seen as a mechanism to avoid trapping and killing the insects that would pollinate them. In fact pollinators and prey are usually from entirely different groups of insects, and the flower height is probably to get them noticed above the relatively flower-poor herbage.

RIGHT
The mosquito wing is not a perfectly smooth transparent membrane, but is covered all over by tiny hairs, thought to help it remain airborne.

LEFT
Mosquitoes are relatively slim insects. Only the females seek a blood meal, to secure enough protein to mature their eggs.

RIGHT
The tipped hairs on a sundew leaf ooze a sweet liquid to attract flying insects, and entrap them in the stickiness.

LEFT
The sundew *Drosera capensis*, a native of South Africa, has caught an aphid in the glistening droplets: now its leaves will curl tightly round the trapped prey.

2

The struggle for existence has led to the development of whole arsenals of weapons, along with tools for feeding, breathing, moving and mating. These complicated and often surreal structures have developed in response to particular needs, and each has a particular function. Hidden from the naked eye until the arrival of high-powered microscopes, these bizarre structures are now visible to be examined in all their complex and otherworldly glory. In this section the structures of the nano world are arranged according to their varied functions.

NANO F

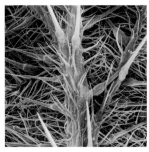

weapons · tubes · claws · suckers
senses · tails · procreation · breathing
hooks · spore bearers

JNCTION

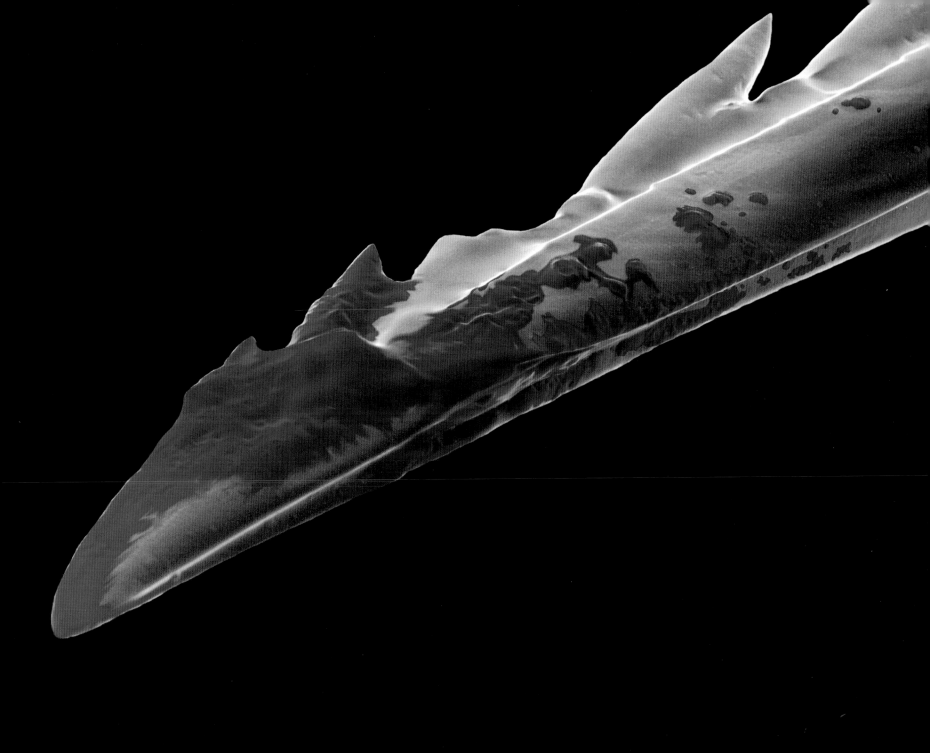

The sting of a honeybee,
barbed to catch firmly into
the skin of the victim.

See page 112

OPPOSITE

Venemous spines covering a stinging nettle leaf.

See page 112

OPPOSITE
Spear-headed bristles protect a carpet beetle from predators.
See page 113

OPPOSITE
Feathery hairs coating the body of a tarantula spider.
See page 113

A weapon, sharp-edged and barbed, menacing with blooded tip – the sting of a honeybee is a frightening device, and has earned the insect respect well beyond its diminutive size. As a weapon, the sting is unsurpassed, and its possession by bees partly explains the deep-seated uneasiness about insects that is felt by so many people.

The most obvious feature of the sting is its barbed edge, and this is also an important aspect of its structure. For as a honeybee stings, it quickly saws into the skin using a rocking motion, and the barbs help open the cut to penetrate deeper. Then, as the venom is injected through the hollow tube, and the victim starts to feel the intense burning pain, the barbs anchor the bee in place and prevent it being easily brushed off. In desperation, the target of the bee's attack may remove the insect, but the sting, ripped from the body of the dislodged bee, remains firmly attached in the skin, and the muscular poison sac continues to pump its chemical pain into the wound.

The angry bee, now disembowelled, will die, but its attack is not over. The action of tearing out its body parts releases alarm pheromones (chemical scents) from the stinging organ, still lodged in the skin. This effectively labels the 'enemy' and recruits other bees from the nest to seek it out and launch a full-scale war against it. Although each bee can deliver only 0.1 mg of venom, the pain generated is a sure sign of its dangerous toxicity. It consists mainly of a compound called mellitin, which destroys red blood cells, activates the allergic reactor histamine and reduces blood pressure. It can take as little as 20 stings to kill an adult human.

This ferocious bundle of dangerous spines is not protecting a cactus, and the pricks they give are not just skin-deep. Each is a venomous needle, and the pain it inflicts is intense and long-lasting.

The stinging nettle is native throughout Europe, Asia, North Africa and North America and is well known as a plant to be avoided. To the naked eye the leaves look downy, with soft hairs, but hidden among them are sharp stylet spines. These spines are hardened with silicate, which also makes them brittle, so when touched they penetrate and snap off. But the pain comes not from the physical puncturing of the skin, but from the powerful venom stored in the swollen spine base, which immediately flows through the hollow needle and into the wound.

The venom contains acetylcholine and serotonin, which are chemical neurotransmitters in humans, as well as histamine, which is associated with the unpleasant side-effects of allergic reactions. The pain from stings is immediate and jabbing, but soon fades to a warm tingling. A few glancing stings produce small pale raised pimples, but a heavier touch may produce an extensive red rash.

Because stinging nettle is so common, there is a wealth of folk remedies against its effects, usually involving rubbing the skin with something cool like saliva, dock or other plant leaves, or using baking soda, which is reputed to counter the tiny amounts of formic acid also found in the venom.

The warming effects of nettle sting have long been noted, and the Romans are sometimes credited with flogging themselves using nettle bunches to heat their bodies in the cold northern climate of Britain. Some rheumatism and arthritis sufferers still use nettle stings to ease their aching discomfort.

RIGHT
The tip of a honeybee's sting is barbed to catch firmly into the skin of the victim. The sting is adapted from part of the egg-laying apparatus, so only females sting.

LEFT
The honeybee colony is virtually all female, and the combined stings of thousands of workers are a good defence against honey robbers.

RIGHT
The sharp spikes of nettle leaves are brittle and hollow, and as they break off in the skin they release a painful venom stored in their slightly bulbous bases.

LEFT
The distinctive saw-toothed leaves are a familiar sight along road verges and on waste ground, wherever the soil is disturbed but nutrient-rich.

Ranks of arrayed spears projecting an impenetrable thicket have long been a standard defence for warring armies. The sharp dangerous tips of the lances keep even the most ferocious of frenzied attacks at bay. And so it is with these bristle spears, which together form the standard defence of a carpet beetle larva against would-be attackers.

As their name suggests, carpet beetles eat carpets, but only those made of animal fibre like wool and silk. They will also turn their attentions to other foods such as silk and wool clothing, fur coats, leather, dried and cured meats, and even stuffed animals in museums. Before humans came along and carpeted their homes, the beetles lived in bird and animal nests, scavenging on fallen feathers and moulted hair and perhaps even the odd dead nestling too. This brought them into close contact with

large dangerous host animals that were more than likely to try and eat them. The long bristly hairs are brittle and break off, leaving the spearhead tips embedded painfully in mouth or beak.

The hairs are also a good defence against smaller predators. The natural home of some carpet beetle species is in the messy tangled spider webs found under the loose bark of large dead trees; here they eke out a meagre existence, eating the remains of dead insects left over by the spiders. Should a spider approach the larva it is met with a thick barrier of spines that break in its jaws.

Carpet beetles have been domestic pests for as long as humans have created dwellings for them to invade, and although the adults are derided as carpet, hide, larder, bacon or bone beetles, the fuzzy larvae are rather affectionately known as 'woolly bears'.

This forest of pine trees looks lush and thick, an almost impenetrable thicket. But these 'pines' are less than half a millimetre high, and the dense thicket they form is the thick coat of hairs on the body of a Mexican red-kneed tarantula.

Tarantulas are large hairy spiders, with leg-spans up to 15 cm. Like all spiders they are venomous, hunting insect prey and killing it using venom injected through fangs on their mouthparts. They are known to bite humans, but despite widespread fear, the bites are very rare and usually about as painful as a wasp sting. Tarantulas are loath to bite people and are remarkably docile – and are therefore hugely popular as fascinating pets.

Biting is a good means of attack, but it is not such a good defence as might at first be supposed. For one thing it is probably too late if the spider has already

been picked up by an aggressive enemy intent on killing or eating it. So tarantulas have another, more subtle, weapon – their bristles. If a tarantula is alarmed, it rears up threateningly, waving its front legs about. While it is doing this it is also rubbing itself with its hind legs, and the hairs that brush off lie on its body or float into the air. Anything getting too close to the spider or picking it up will be quickly contaminated with the bristles.

The hairs are needle-sharp at their tips. In the mouth or beak of a would-be predator, they are painfully gritty and sharp, like tiny splinters. If they get into the eye or soft membranes of the airways they can penetrate deeply, causing serious inflammation and irritation. People who keep tarantulas as pets are warned to wash their hands after handling the animals, and to avoid rubbing their eyes.

RIGHT
The spear-headed bristles of a carpet beetle larva break off in the jaws of any would-be predator, causing painful irritation.

LEFT
The bristles act as an impenetrable barrier or, like this species which scavenges in spider webs, are vibrated to confuse the spiders feeling for vibrations on the silk.

RIGHT
These quill-like hairs, embedded in the skin of a red-kneed tarantula by their pin-sharp tips, are brushed off as a defence by the spider and get into the eyes, nose and mouth of any would-be attacker.

LEFT
Tarantulas make docile pets, but handling them roughly can result in allergic reactions to the airborne bristles.

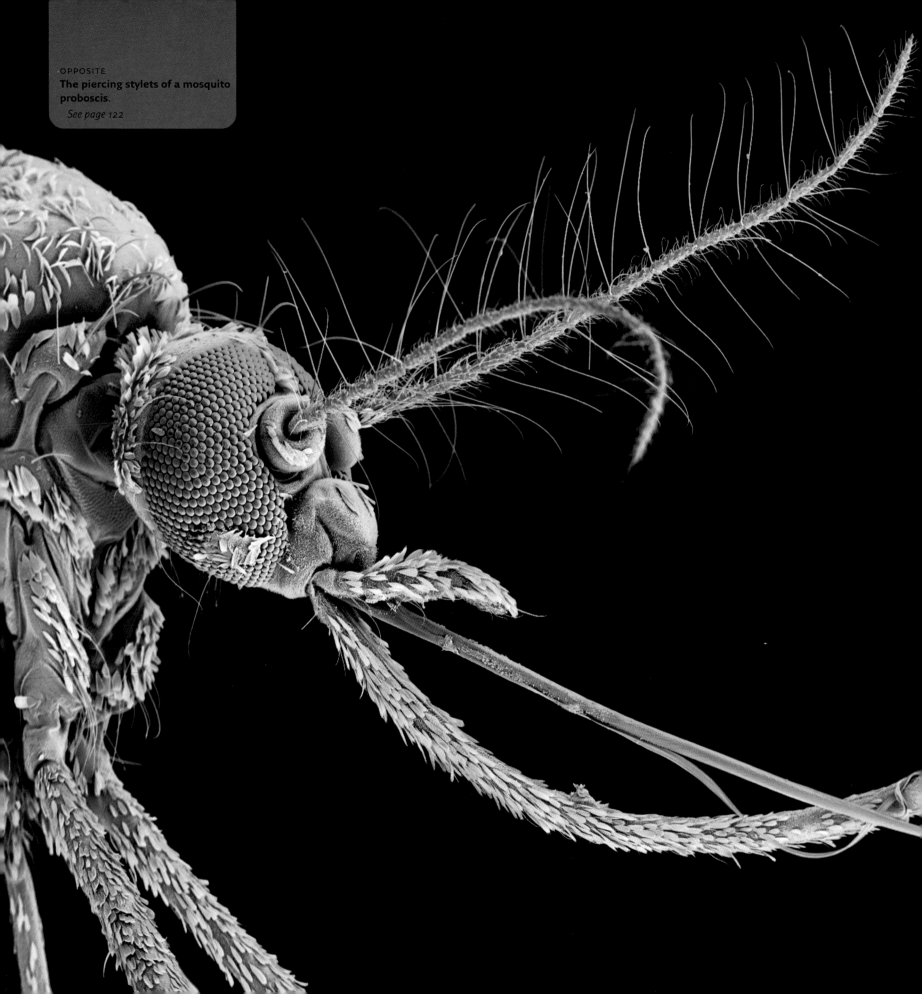

OPPOSITE
The piercing stylets of a mosquito proboscis.
See page 122

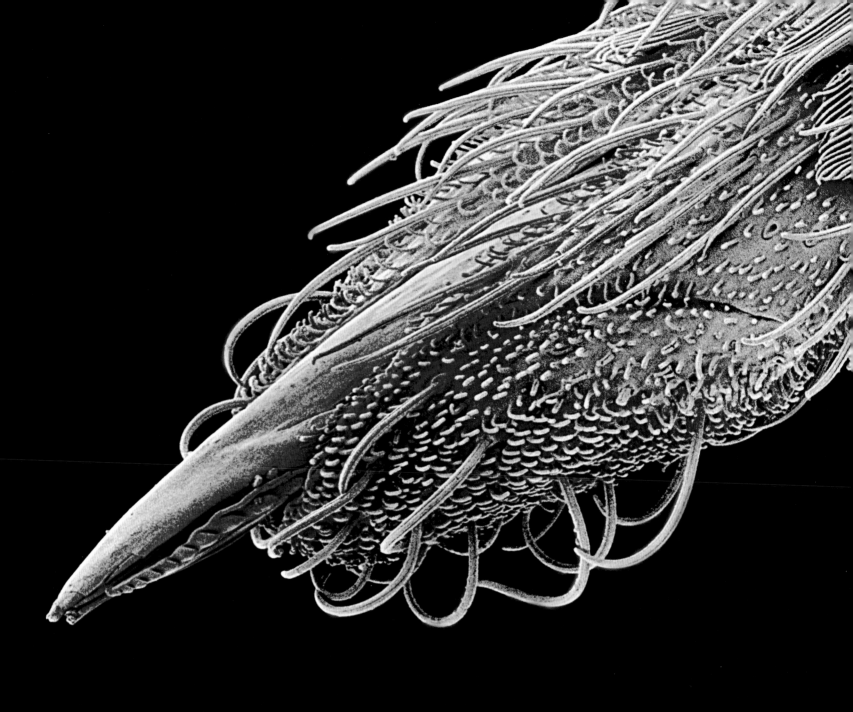

OPPOSITE
A moth proboscis, curled when not in use.
See page 122

OPPOSITE
Section through a bamboo stem, showing the vessels which transport liquids and nutrients.
See page 123

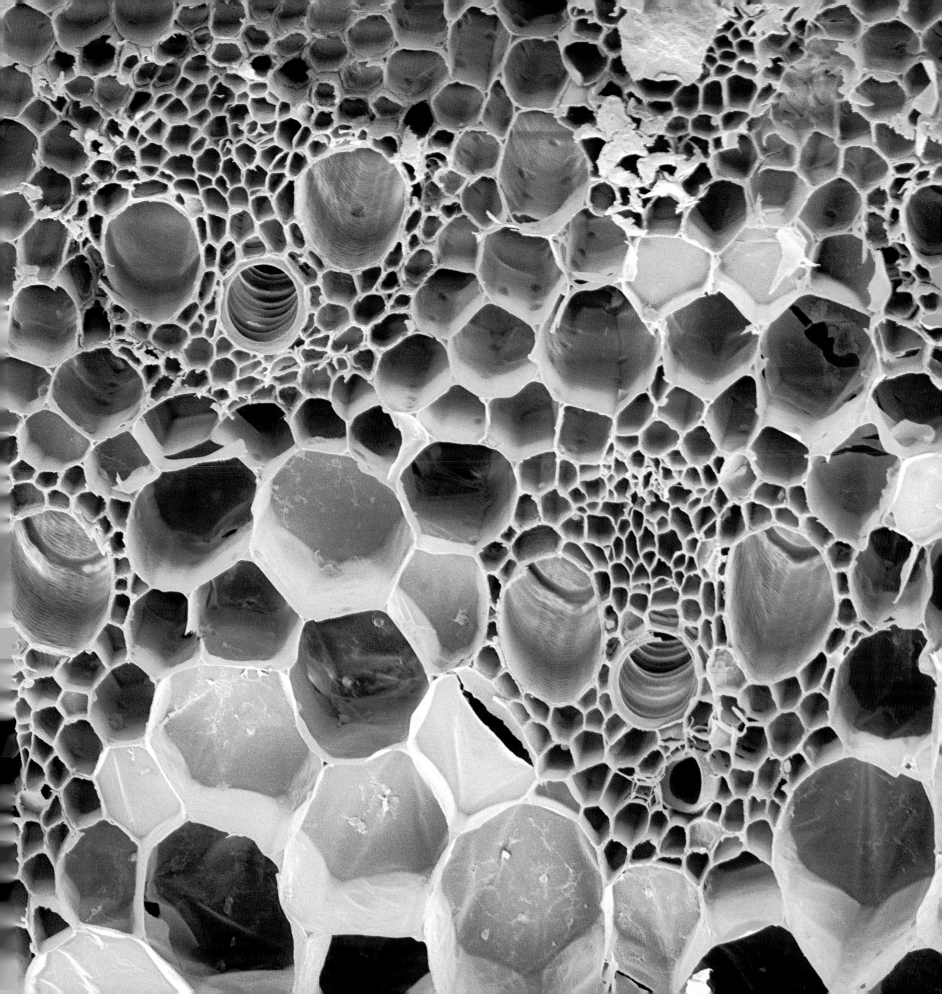

OPPOSITE
Cut guard hairs from the fur of a polar bear.

See page 123

A dangerous weapon, barely sheathed, is intimidating from its slightly curved tip to its smooth glistening upper surface. This image, the stuff of nightmares, shows an object that could have only one purpose – to stab, and to enjoy the flow of blood from the resulting wound. The proboscis of a mosquito is well suited to that single purpose, and is supremely successful in its aim.

As in all insects, the mouthparts of the mosquito are made up of paired pieces, left and right, derived away back in evolutionary history from limb-like protuberances which elsewhere in their bodies have become wings, legs and antennae. Six piercing and sucking elements, together called the fascicle, are wrapped in the labium, shown here as the bristly outer sheath. When the mosquito lands on human (or other animal) skin and presses its mouthparts

down, the labium flexes out of the way. The mandibles, the long smooth outer components that look like upper and lower beak, are the strong penetrating constituents, used to break the first skin barrier. The maxillae, each shorter, more curved and ribbed on the outer side, are the penetrating tools. They cut alternately, sawing the wound deeper and using the rib-like hooks on their outer side to anchor themselves against the flesh.

Hidden inside the cutting implements are further mouthparts: two stylets, fused into a salivary duct, inject enzymes to start the digestion and anticoagulants to prevent blood clotting; two more, fused into the feeding channel, take in the blood meal.

Only female mosquitoes suck blood, using the rich protein to mature their eggs. The males use their stylets to sip nectar and plant juices.

The vertiginous drop into the well of a spiral staircase or the tendril twists of a fire-fighter's hose? It is neither, but the hose image comes closest to the true function of this helix, for these are the coiled loops of a moth's tongue, its proboscis.

Moths and butterflies feed on liquid flower nectar, and use their long sucking tongues to probe the floral nectaries, which are often buried deep inside the protective trumpet of petals. Rather than carry about an ungainly stiff stalk of a straw-like tube, the moth is able to curl up its tongue and store it discreetly under its head until needed.

The tongue is not just a simple elongate hollow cylinder, but a complex device requiring a complex construction. It is made up of two long flexible rods called galeae. These have evolved from two, originally segmented, appendages on the underside of the head. Each rod is grooved on the inner side, and where they meet, along the entire length of the

proboscis, they combine to create a food canal roughly circular in cross section. At the top of the food canal is a muscular pump, which creates a vacuum to suck up the nectar.

The galeae have a natural elasticity, so to extend the tongue to full length the moth increases the internal pressure of the haemolymph (the insect equivalent of blood), causing it to erect and stiffen hydraulically. To retract it, the moth relaxes the pressure and then contracts a series of short oblique muscle strands that line the galeae, bringing the tongue into a tight stowage curl.

This ability to dramatically increase the length of the tongue means that moths can sip nectar from the deepest of flowers, where only they can reach. In some instances a moth's tongue is many times the length of its body, a feature that has evolved in parallel with a hugely elongated flower, so that a particular moth species is its only insect pollinator.

RIGHT
When mosquitos bite they also inject part of their last meal, including the protozoan parasites, *Plasmodium*, which cause malaria.

LEFT
As the flexible protective sheath is folded back it reveals the mosquito's long blood-sucking mouthparts.

RIGHT
The tight curl of a moth's tongue allows it to stow its nectar-sucking organ, which is often many times longer than its body.

LEFT
The hummingbird hawkmoth hovers at a flower whilst extending its extremely long proboscis deep inside.

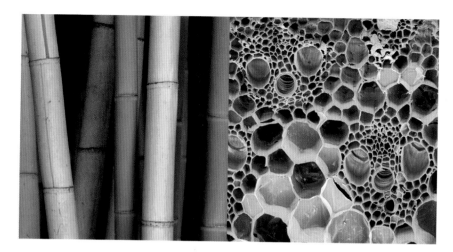

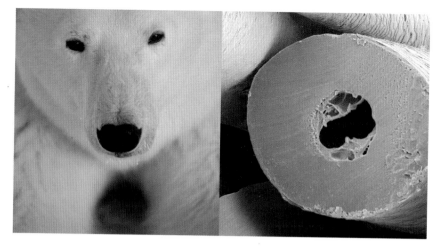

This could be a stack of drinking straws packed tight together. The resemblance is no coincidence, because these tubes, running lengthwise through the stem of a bamboo, carry out exactly the same hydraulic function – the controlled movement of fluids under pressure.

Each group of tubules, coloured lime-green in this image, is called a vascular bundle, which gives vascular plants their name. This huge group includes ferns, flowering plants and conifers – in other words the majority of green plants. Non-vascular plants include much simpler vegetable forms like mosses, liverworts and green algae.

A bundle contains two types of tube. The larger cells in each group are xylem, which carries water and absorbed mineral nutrients from the roots up to the leaves. There is no energy expenditure on the part of the plant, which moves the water by a combination of water pressure pushing in at the roots by osmosis and evaporation from the leaves sucking the water upwards. Xylem tubes are tough and fibrous; they are usually no longer growing, containing no working cell contents like nucleus or cytoplasm, and they are the main constituent of wood.

The smaller cells make up the tissue phloem, which carries photosynthesized nutrients in the form of sap sugars away from the leaves, to other growing parts of the plant or to storage in the roots, bulbs or corm. Phloem cells are shorter than xylem, but are joined to each other into tubules by sieve-like connection plates at each end.

Throughout the plant kingdom, xylem and phloem are arranged in the same way, with the woody xylem nearer the centre of stems and roots, phloem on the outside. In leaves xylem runs closer to the upper surface, with phloem below it. This feature is exploited by aphids, which usually feed by sucking the sweet sap from the underside of leaves.

This could be something as domestic as a stack of toilet rolls or kitchen towel. Even the shear pattern, where it has been cut through, makes it look as though the soft spongy paper has been compressed and torn. But this image is very far from household domesticity, because these hollow tubes are the hairs from a polar bear.

These are the guard hairs, the coarse outer layer that protects the shorter and finer underfur from the elements. Although appearing white to the naked eye, polar bear fur is translucent. It was once suggested that the hollow stems acted like fibreoptic tubes, taking light down to the bear's dark skin, but it now seems more likely that the air-filled hollow simply aids insulation. The bear needs all the heat-proofing it can get, since it has adopted one of the most inhospitable landscapes on earth for its home – the frozen Arctic. Another Arctic mammal, the reindeer, also has hollow guard hairs.

The bear spends virtually all of its life on the ice, or at least on the edge of the frozen ocean, where it fishes for seals and other prey in cracks and in the sea. Any other fur would make it more obvious, so although it moults in summer it never changes to a darker colour like many other Arctic creatures. In captivity, if kept in warm humid conditions, polar bears sometimes turn a pale green, as algae grow inside the hollow hairs.

The guard hairs vary from 5 to 15 cm in length over most of the bear's body, except on the front legs of the males. Here they grow into a long dense mane, up to 35 cm long, and this is thought to act as a sexual signal to females.

RIGHT
Bamboo is a tall and fast-growing plant, and it needs to move fluids quickly and easily through its stems.

LEFT
A section through a bamboo shoot reveals the complex hexagonal array of tubules that carry water from the roots to the leaves, and photosynthesized nutrients from leaf to growing stem.

RIGHT
Hollow hairs help to insulate the polar bear against the cold.

LEFT
The bear looks white to the naked eye, but the colour is the result of light diffraction, for the hairs themselves are not pigmented, but translucent.

OPPOSITE
**Twin claws from the foot of a
hedgehog flea.**
See page 132

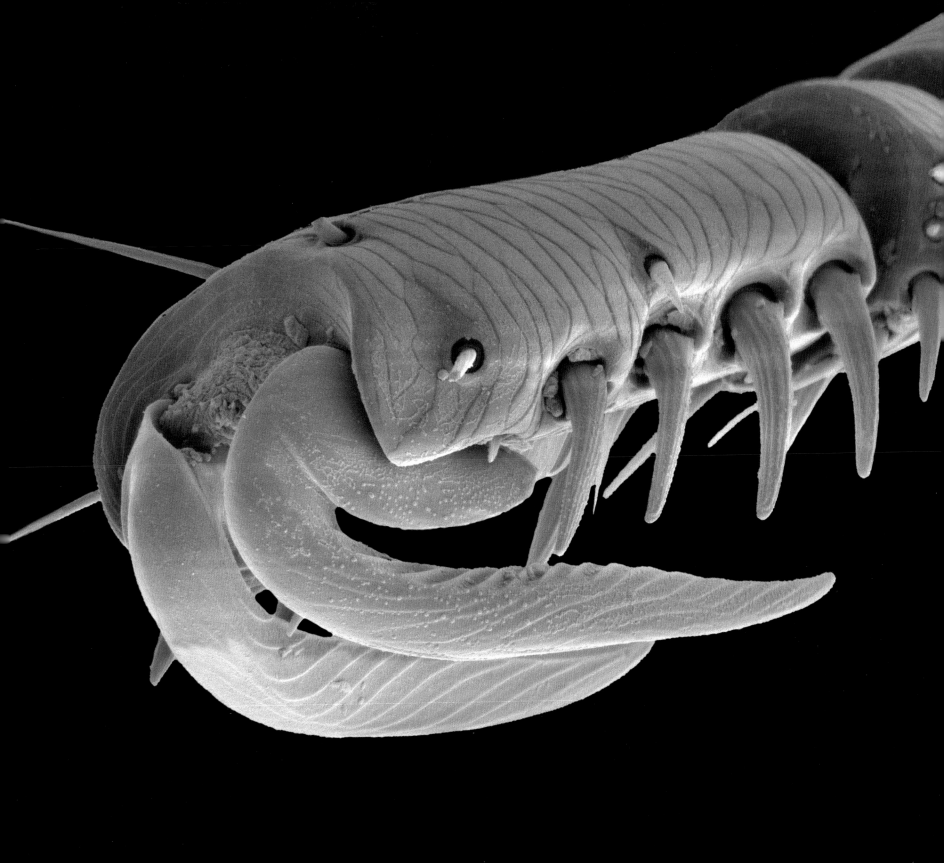

OPPOSITE
Claws from the foot of the Hercules beetle.
See page 132

OPPOSITE
Hair-gripping claw of the human head louse.
See page 133

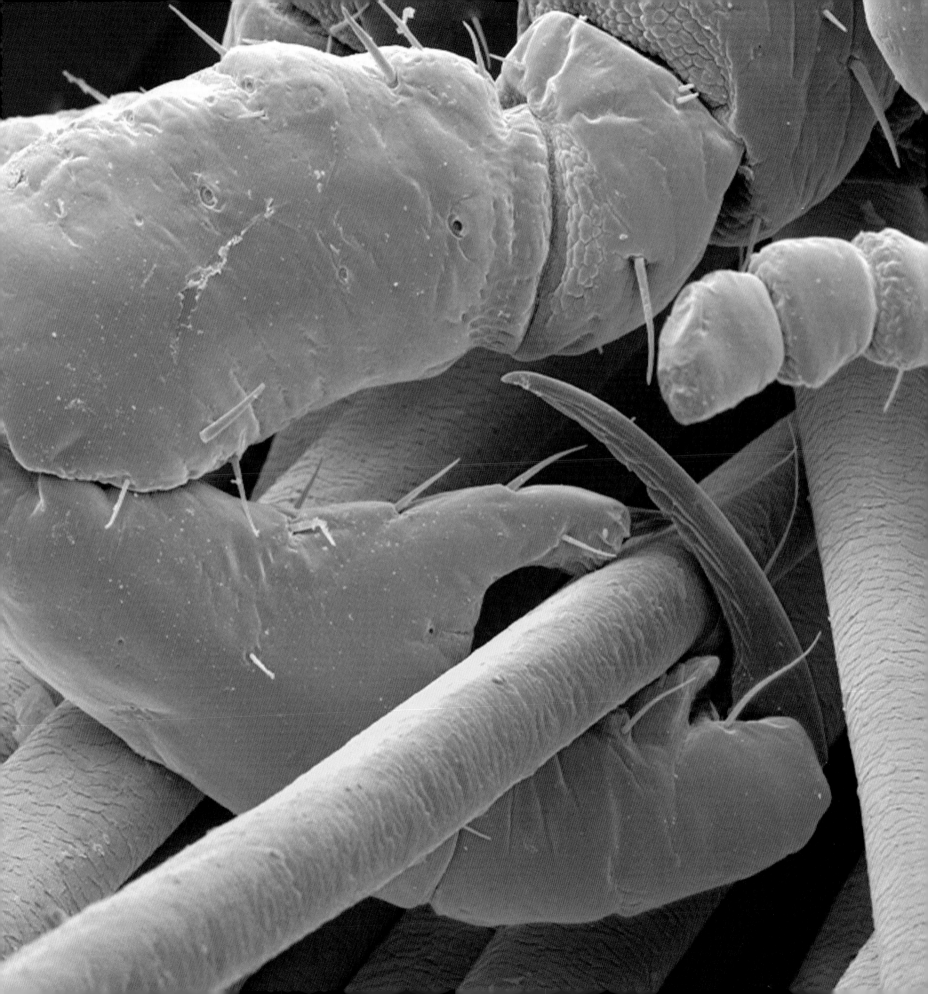

OPPOSITE
**Claws, pegs and hairs at the tip
of a spider's foot.**

See page 133

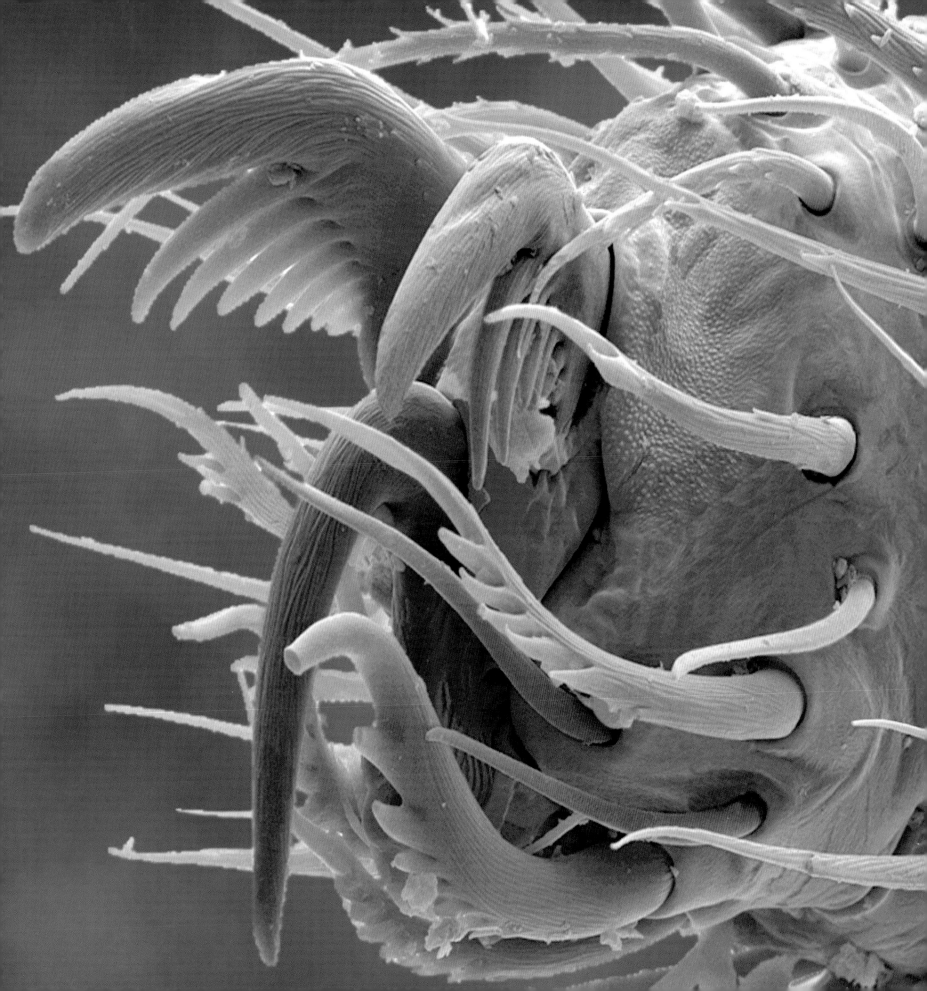

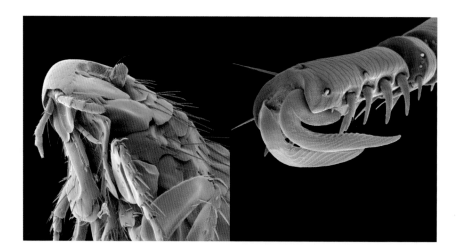

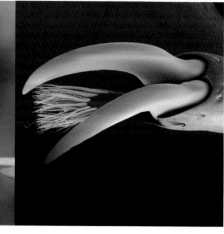

This powerful grappling iron has a double hook to take extra weight, and restraining pins to prevent cable slippage, but is never hoisted on a dockside cargo crane. Nor is it a new form of rock-climbing karabiner, although it looks as if it could secure a whole party of climbers and their tents in a ferocious gale. Instead these hooklets are used for a more delicate grip, but one no less important for their owner – the hedgehog flea.

Hedgehogs are renowned for their sharp spines, each a large hollow hardened hair, which they use for defence. But underneath the quills is a thick pelt of normal hair, and it is in this that the hedgehog flea lives, sucking the blood of its unfortunate host. Fleas, like other fur- or feather-dwelling parasites, are constantly in danger of being removed by grooming, scratching or moulting, so they have evolved stout claws to cling on tight. The recurved

hairpin claws at the tip of the flea's tarsus (foot) articulate, and close against the row of four sharp pegs on each side, trapping the hair strands between them.

The gripping action of the flea's claw is also important when it comes to the insect's jump. The claws give firm purchase on the ground as they push off. Fleas jump by building up huge elastic tension in the muscles of their fat hind legs, then suddenly releasing it. The massive flick sends the insect shooting into the air, making a leap 100 or more times its own body length.

Fleas have another feature that allows them ease of movement on their hosts, so they almost swim through the fur. They are very flattened, and the head, thorax and each of the abdominal segments are all armed with a comb of strong backwardly projecting spines. They can readily move forwards, but cannot easily be dislodged backwards.

The smooth-textured surface gives these two appendages the appearance of curved ivory tusks, but they are no teeth, or weapons – they are claws, and their purpose is to grip. These are the claws at the end of a Hercules beetle foot.

The Hercules beetle is a large and powerfully built scarab. It is found in the tropical rainforests of Central and South America and has long been a prize for collectors and museums for its beautiful and strange form. It is one of the largest rhinoceros beetles, up to 17 cm in length, and the male possesses two long narrow horns, one growing forward and down from the centre of its thorax and one growing out and upwards from the centre of its head. They meet, almost like pincers, and are used in jousts between rival males fighting over females.

The beetle's body is thickset and its legs are short, so wrestling struggles are

decided on the results of leverage as a victor finally grasps and topples its defeated opponent. Since they breed in dead and decaying timber, battles usually take place on logs and tree trunks, where a firm grip on the crumbling bark substrate is of paramount importance. The two claws on each tarsus (foot) are large and powerful to anchor the beetle in its territorial competition.

Almost all insects have two claws on each foot, a shared character illustrating their evolutionary descent from the same common ancestor hundreds of millions of years ago. The central tuft of hairs showing between the claws is called the arolium. In insects like flies, this pad is greatly expanded and flattened and acts as a means of adhering to hard smooth surfaces (see the fly's foot on page 37). The Hercules beetle never comes across such situations, instead relying on its stout tarsal claws to hold it in place.

RIGHT
Almost all insects have two claws at the end of each leg, curved for various gripping strategies; the flea has a hairpin bend on each, to grip firmly around the hair of its host.

LEFT
Fleas are adapted to life pushing through fur, with flattened bodies and rows of backward-pointing spines.

RIGHT
The two claws at the tip of the beetle's foot grip firmly onto logs and tree trunks as the insect hauls its huge bulk about the tropical rainforest.

LEFT
The Hercules beetle is so-called because it is a large and powerful insect.

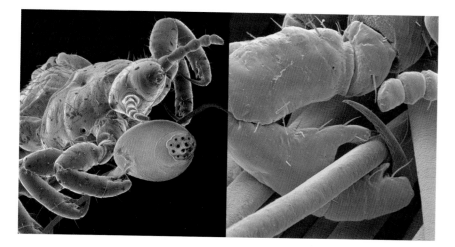

This must rank as one of the best gripping devices of the animal world. The sheer size of the giant muscular hand, the sculpted groove to receive its tubular cargo, and the massive angled finger to clamp the claw shut, all testify to its immense power and strength. And grip it does, because holding on is a matter of life and death to a head louse: if it lets go, it's dead.

Head lice are tenacious bloodsucking parasites that have tormented humans from before recorded history. They move through the scalp hairs with remarkable ease and speed given the ungainly size of their feet. If heads brush against each other, a louse is able to scamper quickly onto its new host in a matter of seconds. It cannot jump or hop or float, but it can crawl very fast.

Only death, either its own or that of its host, will ever induce a head louse to relinquish its grip. The louse has no reason to let go: it has the perfect habitat and an easy life. The human scalp is warm and humid and safe. The louse's food, blood, is readily available whenever it chooses (usually about five times a day), and with luck the host human will keep the habitat clean by washing its hair regularly and preventing any build-up of unpleasant louse excrement.

Lice breed rapidly, attaching their tough grey eggs, called nits, to the base of hairs, hard up against the scalp. On hatching, the tiny louse nymph leaves behind a bright white egg-shell, distracting its host to uselessly pick up the empties.

If combed out from the hair, the louse is doomed. Away from the moist skin of its host it will quickly dehydrate and die within hours. No wonder it holds on so very tight.

With its selection of variously toothed blades and hooks, this fearsome array of weaponry would not look out of place on a Swiss army knife. Perhaps that's not too surprising, given that this is a complex mechanism for a complex and dangerous task. These hairs, pegs and combs are not offensive though, they are defensive – a defence against a spider being tangled in its own web, because they are the claws at the very tip of a spider's leg.

Technically, spiders do not have feet, in the sense of having an angled joint in the limb and a flat pad to rest on the ground. They are ungulate, meaning they walk on the tips of their legs, much as hoofed mammals walk on tiptoe. When a spider walks on a leaf, or climbs up a stem, it uses the uppermost two large long-toothed claws (coloured pale grey here) to grip its way. This is the garden spider, an orb-web spinner, and it spends much of its time walking around its sticky silk snare. The remaining claws are for this tightrope living.

When the tip of the spider's leg is pushed up against a silk line, the stout hooked claw (brown) flexes back, allowing the strand to rest across the smaller upcurved short-toothed hairs. These hairs are elastic and bend backwards, now allowing the stout claw to close over the top of the silk and clasp it firmly against them. Rather than using a broad grasp, like a human hand holding a thick rope, the spider is holding itself by using a tight pincer-like grip on a very narrow point on the sticky silk. When it wishes to let go, the gripper claw lifts up and the flexible hairs spring back, flicking the silk wire away.

RIGHT
The clawed feet of the head louse are supremely adapted for clinging to individual hair shafts. Rigorous combing is the best way to remove these troublesome parasites.

LEFT
A head louse clings tenaciously to a human hair, along with its egg, a nit, showing its breathing holes.

RIGHT
The hooks and claws at the end of a spider's leg allow it to walk on hard surfaces like leaves and walls, but also to grip carefully to its silk web without entangling itself.

LEFT
The silk of a spider web is a complex spiral of sticky strands attached to a supporting structure of radiating spokes.

OPPOSITE
**A caterpillar's stem-
gripping proleg.**
See page 142

OPPOSITE
Rows of suckers along the arm of an octopus.
See page 142

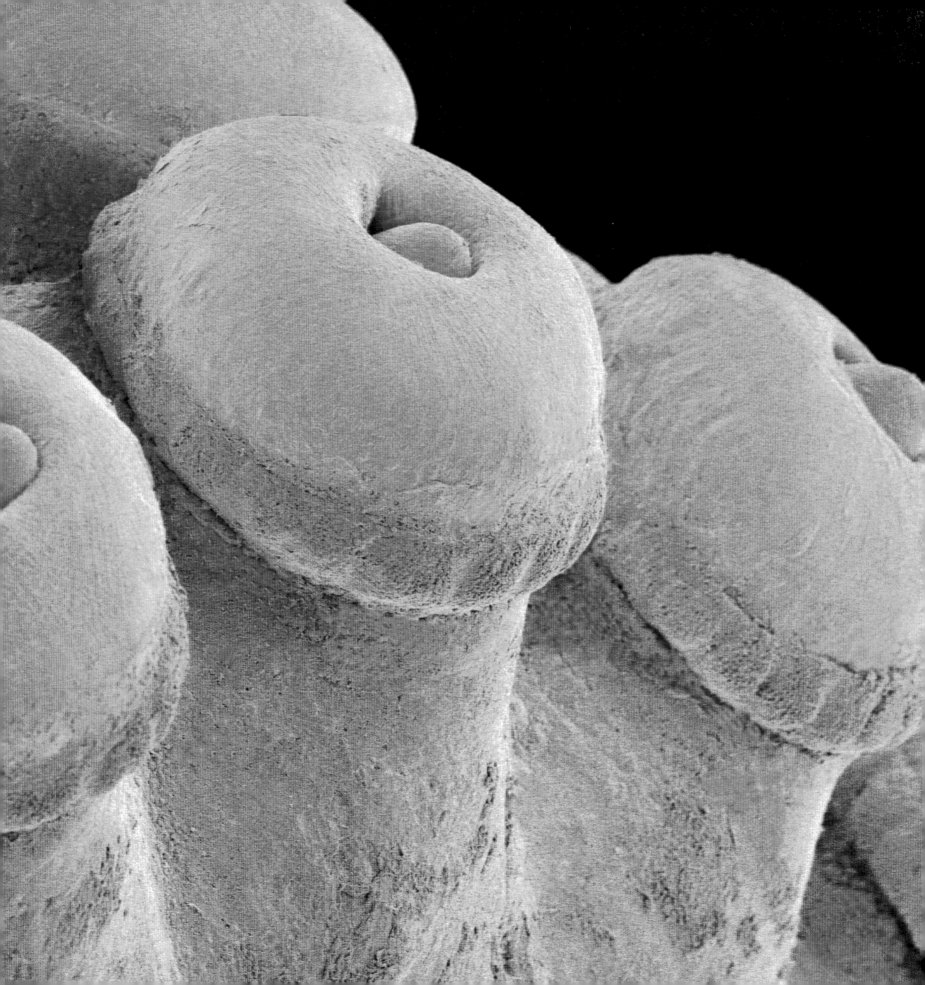

OPPOSITE
**Head and sucker of a male
schistosome fluke.**

See page 143

Tail sucker of a medicinal leech:
the animal's main grasping organ.
See page 143

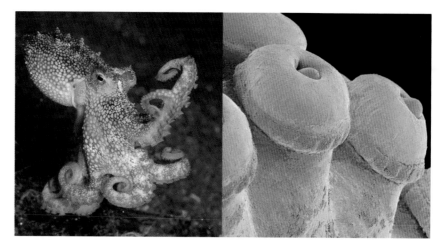

The yawning hook-toothed sucker mouth of some alien worm appears to be reaching up, hungry for its next meal. But this orifice is simply the empty inside of a circular caterpillar foot, and the claws are its means of holding onto a leaf.

Moth and butterfly caterpillars have a variety of limbs with which they cling onto their food-plants. Although apparently just a uniformly cylindrical eating machine, a caterpillar is already divided into the usual insect body parts: head, thorax and abdomen. Each of the three thoracic segments has a pair of short stiff claw-like true legs. These will eventually metamorphose into the jointed legs of the adult insect, but in the caterpillar they are mainly used to hold and manipulate the leaf as it is presented to the chewing jaws. Of the ten abdominal segments, three or four will have pairs of the soft sucker-like feet, called prolegs, seen here. There is also usually a terminal pair, called claspers, on the final segment. These short fat feet are the main clinging limbs of the larval

stage, supporting the torso of the caterpillar and holding it onto the stem or leaf on which it feeds. Combined with the worm-like flexibility of the caterpillar's soft convoluted body (the folded skin is easily visible here), the relatively simple structure of the prolegs is perfectly sufficient for manoeuvring the animal about on the plant. Despite this important task, they will completely disappear when transformation into adulthood occurs.

Somewhere along the evolutionary history of living organisms, simple animals became more complex by repeating themselves and developing multiple body segments. Each element had its own muscles, nerves, breathing tubes, and its own limbs. Over time, some of these have remained units of locomotion, others have evolved into antennae, palps, wings and tail structures, some have disappeared altogether. The prolegs of caterpillars are embryological echoes of a time when more body segments had appendages.

This row of soft pale mounds ranged on a series of short stalks looks like some sort of baker's confection. It's certainly edible, but not sweet, because these are suckers on the arm of an octopus.

A double row of suckers adorns each of the octopus's eight arms – these are not tentacles, like those of squid and cuttlefish, which are longer and have suckers only at the tips. Each sucker is a soft, flexible, muscular node, and unlike those terminal squid suckers they are not armed with hooks. The suckers give the octopus enormous dexterity for manoeuvring itself across rocks or oceanic plants, or when holding its prey.

The mechanism of each sucker is much more complicated than a simple rubbery cup, like those attached to toy arrows. Each can exert its own suction pressure, or release, to cope with an infinite variety of objects or surfaces.

The main 'foot' of the sucker, a soft ring of muscular tissue called the infundibulum, is relaxed and appears

slightly bloated here. When the octopus is active it is usually held in, to give a slightly concave face. Its flexibility allows it to make complete watertight contact with virtually any surface it touches against. Inside the pore opening is an open cavity called the acetabulum, which is also lined with musculature. When a sucker is placed onto a surface, radiating muscles within the infundibulum contract, stretching the interior cavity. This reduces the water pressure within the chamber, effectively creating suction, and the sucker is pulled down onto whatever it is touching. Circular muscles inside the sucker contract to reverse the water pressure differential, and the sucker is released.

Octopuses are the most intelligent invertebrates known, and they are able to learn elaborate actions such as moving through mazes or opening screw-top jars of food. Despite this, their brains are not complex enough to hold a mental image of their world or of what they are holding in their supremely tactile limbs.

RIGHT
The soft sucker-like foot of a moth caterpillar is able to grip into the surface of a leaf using the multiple rings of curved hooks called crochets.

LEFT
Rearing up on its 'extra' rear legs allows the defenceless death's head hawkmoth caterpillar to look menacing and dangerous.

RIGHT
Each sucker in the double row down the arm of an octopus is under its own muscular control, giving the animal astonishingly dextrous mastery over its limbs.

LEFT
An octopus uses its suckers to creep along the sea bed, to move rocks to create a shelter, or to grasp its prey.

Here is marital togetherness at its most bizarre, as two trematode worms (or flukes) live virtually one inside the other for their entire adult lives. The smaller 'snout' (coloured pale orange in the right-hand picture) is the head of a female *Schistosoma*, peeking out from a groove, called the gynaecophoric channel, on the underside of the larger male. Their close proximity allows fertilization of the many eggs released during their parasitic residence in the human liver, a condition called schistosomiasis or bilharzia.

Schistosomiasis is a common disease in Asia, Africa and South America, with water snails acting as intermediate hosts. Eggs released in human urine or faeces hatch in fresh water into microscopic free-swimming larvae called miracidia, which search for and then penetrate the aquatic snails. Here they live the first part of their life cycle, feeding inside the snail and producing new larval parasites called cercariae.

The cercariae emerge from the secondary snail host in response to the turbulence and shadows caused by someone walking or swimming in the water nearby. Like the miracidia, cercariae are swimmers and actively seek out human skin. They penetrate, usually in the foot or lower leg, and shed their forked swimming tail. The tiny organism moves into the surface veins of the skin, and travels around the body in the blood until, after about 10 days, it reaches the liver, bladder, intestines or rectum. Using its sucker, a worm attaches itself to the wall of the vein and starts to feed on red blood cells. It takes about 45 days for the worms to reach adult maturity. They then pair up and start releasing 300–3,000 eggs per day.

Bilharzia afflicts 200 million people worldwide. The symptoms (abdominal pain, diarrhoea, fatigue, fever) are caused not by the worms but by the body's chronic immune reaction to the eggs that do not escape and which become lodged inside the abdominal organs.

It is obvious that this convoluted corrugated cup has only one purpose: it is instantly recognizable as a sucker, and it is meant to stick to things. For a leech, sticking to things is a matter of life and death, because this sucker is how a leech holds onto its prey.

Leeches are related to earthworms, and have long segmented bodies. Many have adapted to life in fresh water, but in wet rainforests land leeches live on the forest leaves, and in drier habitats they live in moist soil. All leeches have the same number of segments, 34, all more or less identical except for the sucker at each end. The small sucker at the head holds the mouthparts against the prey whilst the jaws (or in some species a piercing stylet) penetrate the skin and suck out the blood. The large sucker at the tail end is much bigger and stronger, for it must anchor the whole animal on its victim even when the leech's copious blood meal increases its body size many times.

Different leeches attack different classes of prey: invertebrates (including worms and even other leeches), fish, birds, amphibians and mammals. The medicinal leech shown here specializes in attacking mammals, particularly livestock, and this brought them into contact with humans.

Despite the squeamish abhorrence often associated with sanguivory (blood sucking), leeches have been used in medicine for over 2,000 years as a means of blood-letting. Originally this was seen as a means of reducing illness by ridding the body of some of its 'bad' blood. Such was the desire for the treatment that, at its height in the mid-nineteenth century, leeches were in short supply because they were so widely exploited. By an ironic quirk, leech blood-letting has made a comeback, and these strange animals are still used in modern medicine – to reduce swelling and bruising to delicate tissues, especially after plastic surgery.

RIGHT
Anchored by the male's sucker, a pair of *Schistosoma* flukes feed on red blood cells in the veins of the lower abdomen.

LEFT
The female of the human bilharzial worm lives inside a cleft on the underside of the larger male.

RIGHT
The wrinkled skin in and around the tail sucker of a medicinal leech is a clear indication of the rubbery flexibility required of this organ, used to attach the animal firmly onto its hapless prey.

LEFT
The leech also uses its suckers to transport itself around.

OPPOSITE
**Ball-joint of the antennal socket
on the head of an ant.**
See page 152

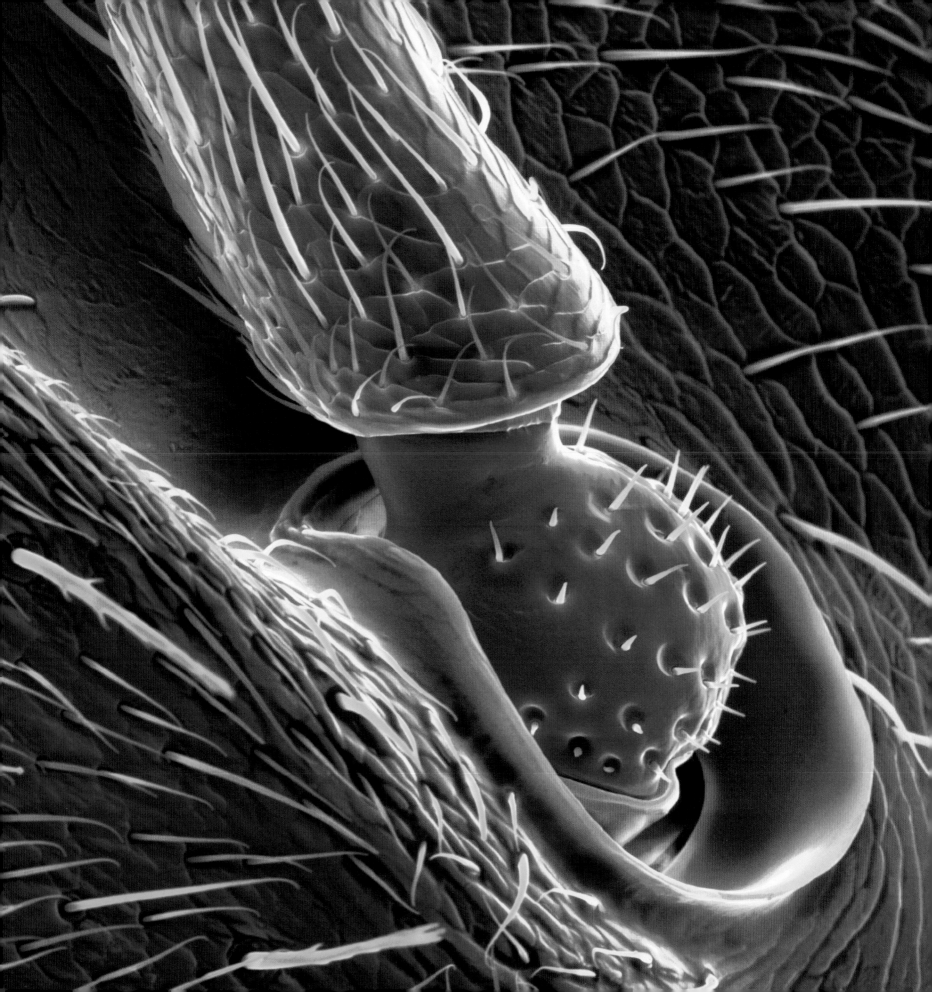

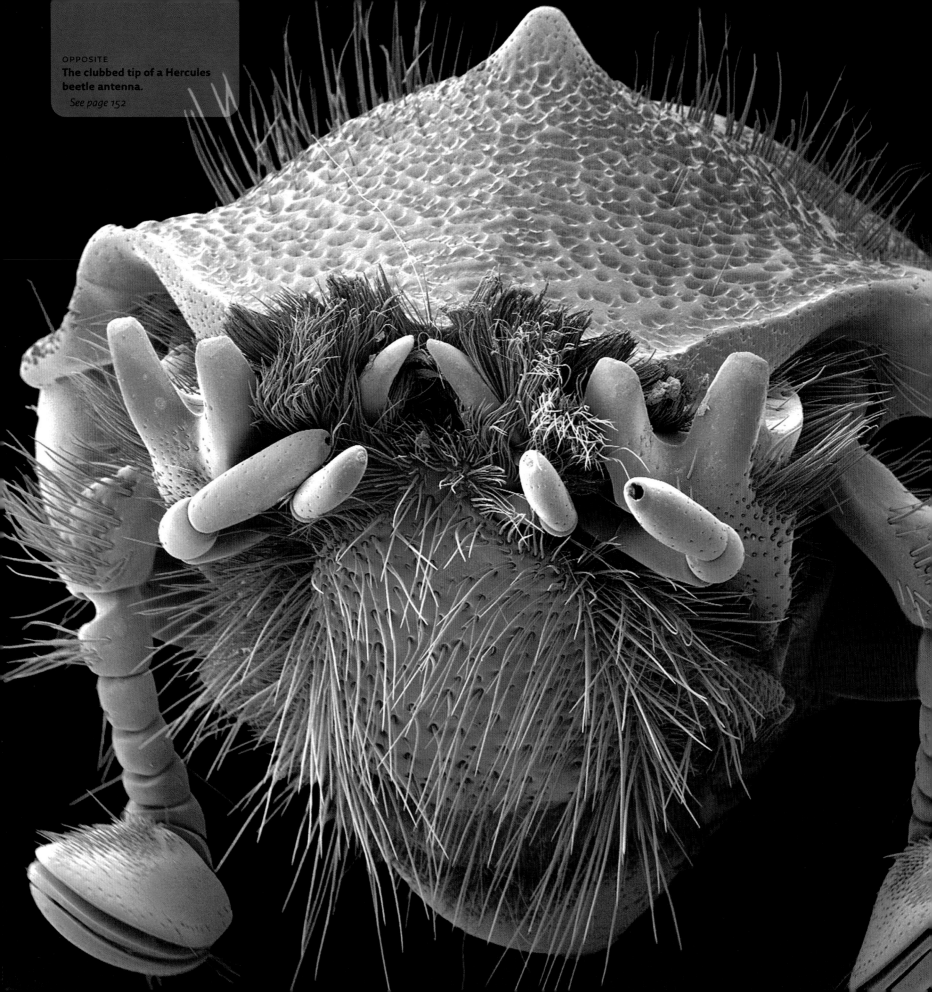

OPPOSITE
The clubbed tip of a Hercules beetle antenna.
See page 152

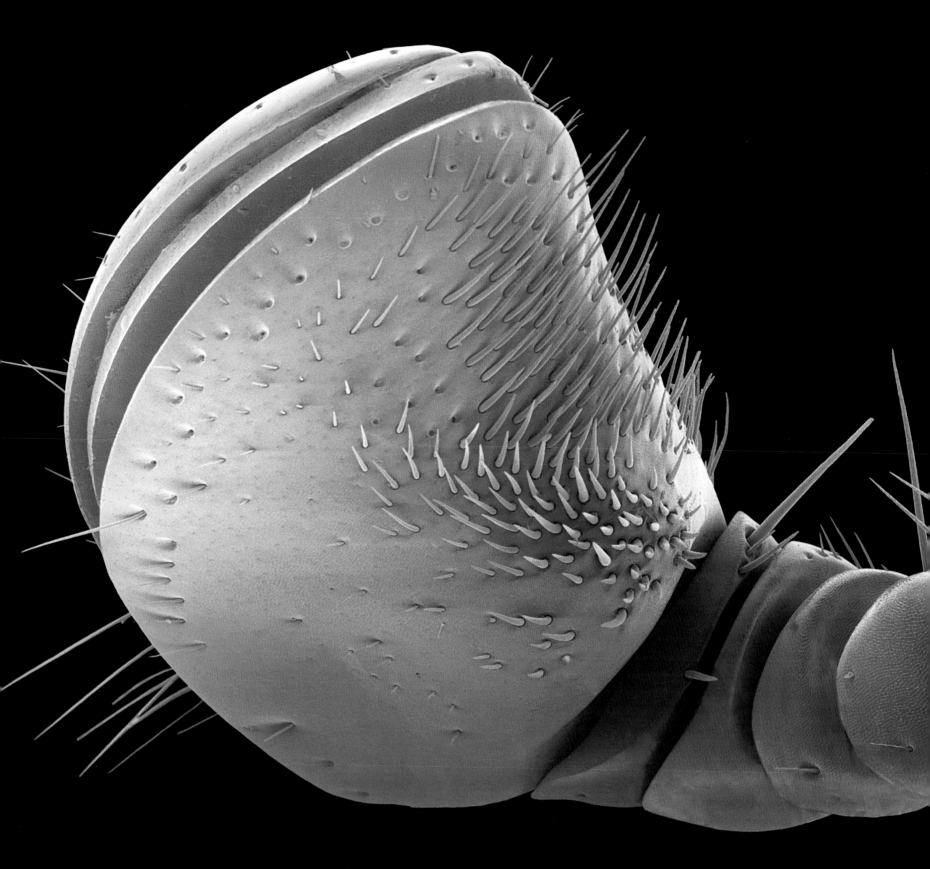

OPPOSITE
Sensory hairs and pits on the antenna of a moth.
See page 153

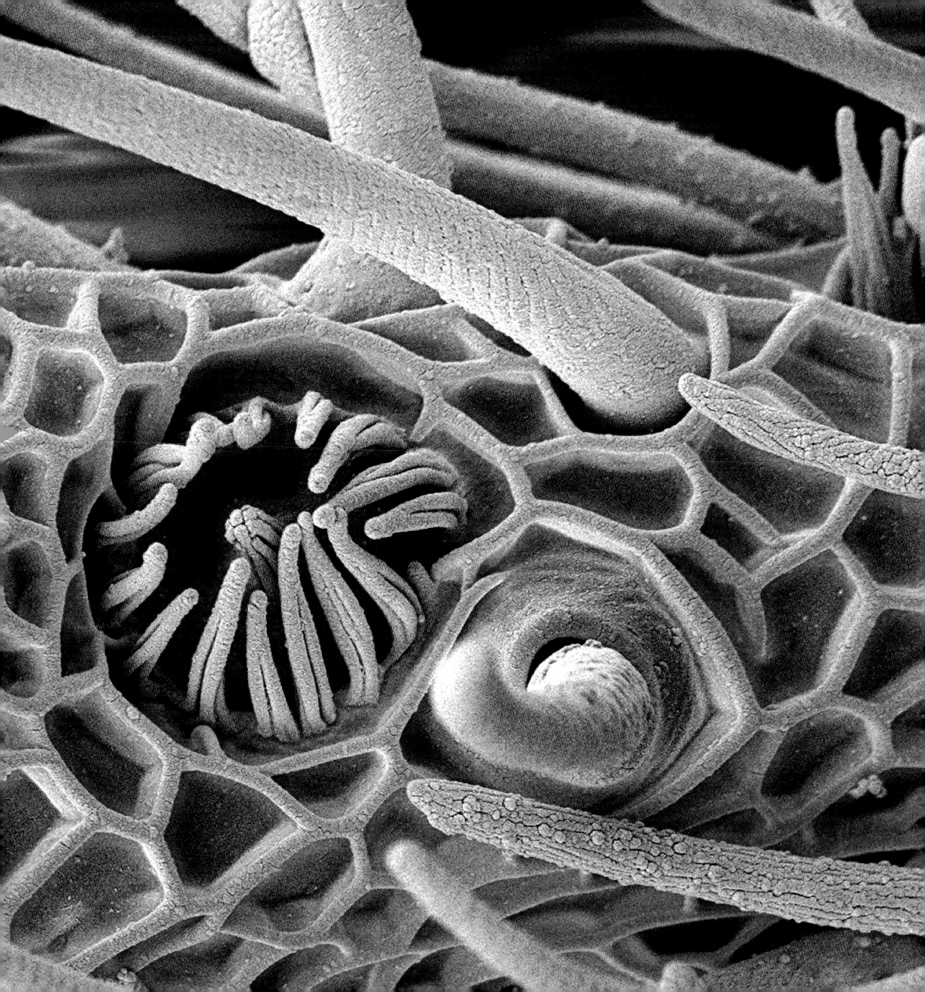

OPPOSITE
The haltere, or balancing organ on the body of a fly.
See page 153

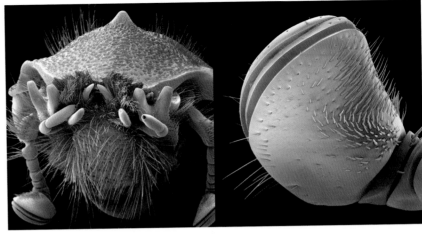

The ball and socket joint has arisen many times in nature. Perhaps the most familiar will be the hip and shoulder joints of mammals, transformed to leg and wing joints in birds. The success of this hinge is in its free rotational movement, allowing the limb to swivel unencumbered through any axis. In the invertebrate world, ball and socket joints occur where legs are attached to bodies and, more importantly, where antennae (feelers) are attached to the head.

Insect antennae are supremely important sense organs, used to detect chemical scents, vibrations and air movements. In ants, they are also used as a means of tactile communication, with members of a colony touching and caressing each other to feel and smell who it is they are confronting.

Male ants usually have 12 antennal segments, while the females (queens and workers) normally have 13. The first segment, called the scape, is long and cylindrical, and often makes up half the total length of the antenna. The ball at its base is connected to the socket, called the antennifer, on the front of the head. Attached to the end of the scape is a short conical segment called the pedicel, followed by the even shorter segments that together make up the flexible part of the antenna, the flagellum. As in all insects, only the first two segments, scape and pedicel, have controllable muscles to move the antenna about.

In ants there is a distinctive angle, of about 90 degrees, between the long scape and the rest of the segments, giving the antenna a characteristic elbowed form. This gives the ant a broad field of manoeuvrability through which it can accurately and deftly sweep its feelers.

With its flexible jointed stem adjoining a striking head of laminated plates and seemingly aerodynamic pins, could this be a new design for a high-tech golf club? Its sheen looks almost machined, and its combination of sharp angles and bulbous lobes looks almost robotic. Its bold form and stout construction belie its delicate use, for this is an antenna from a beetle.

The Hercules beetle is part of a large and successful group, the Lamellicornia, found throughout the world. They feed on plant material, and although a few have become minor horticultural pests because of their nibbles on garden plants, most feed on decaying vegetable matter, either in rotten wood or in plants which have already been eaten by herbivores and passed out as dung.

They are characterised by their club-like antennae, where the terminal three to five segments have become expanded into broad flattened plates called lamellae. The size of the terminal plates varies between species, but is usually largest in the males, suggesting that one of the functions of the enlarged segments is to detect females from their pheromones (sexual scents).

This also borne out by studies on the sensillae (minute chemical receptor organs), which cover the antennal segments. Particular types of sensillae occur only in slight indentations on the inner sides of the lamellae, and they are usually much greater in number in males than in females. By electronically testing individual receptor nerves in these sensillae, it has also been shown that although some are sensitive to plant chemicals (so the beetles can smell out their foodstuffs), others react only to the very specific airborne chemicals produced by the females of the species, and they react to minute concentrations, just a few molecules firing off a nervous response.

RIGHT
The long basal segment, combined with the multisegmented flagellum, makes the antenna the insect's most important tactile sense organ.

LEFT
An ant's antenna is connected to the head with a ball and socket joint, offering it universal movement.

RIGHT
The club of a Hercules beetle's antennae contains chemical scent detectors that help the beetle smell out its food, and females.

LEFT
Although they are important sense organs, the antennae can be tucked away under the head as the beetle roots in soil and leaf litter.

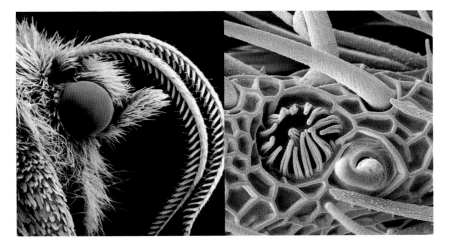

The soft flexible-looking triple tendrils rising in a ring around a darkened hollow could almost be a sea anemone embedded in a shaft of latticed coral. But this is the surface of a single cylindrical segment of a moth antenna, and the hairs, fronds and dimples are part of the complex sensory system the moth uses to navigate its way around the world.

Moths have highly developed antennae, which are much more than simple mechanical feelers. They are the organs of smell and taste as well as touch. The hairs, bristles, pits and hollows are different sense receptors, and each moth species has its own unique arrangement for its own unique needs.

Six or seven distinct types of detecting structures, called sensillae, are found on almost every moth antenna studied, each defined by its external shape and the internal nerve system. The sea anemone arrangement is a fence of projections surrounding a sprig of sensory hairs used for smelling – useful for finding nectar sources and larval food-plants. The palisade ring of protrusions slows down the air moving over the chemoreceptors, allowing a more detailed analysis.

The commonest are the large hair-like strands, here coloured pale pink, which cover most of the antennal surface. Called a sensillum trichodeum (meaning, very sensibly, 'hair-like sense organ'), each is covered with an array of minute pores arranged in gentle spiral grooves.

These are the sensors responsible for detecting the minute amounts of pheromones, sex scents, released by female moths to attract mates. A single airborne molecule of these complex chemicals is enough to trigger a nerve impulse in the male antenna. To increase the pheromone detection, male moths often have large feathery antennae, greatly increasing the surface area on which large numbers of sensillae can be packed.

This simple hairy bean-shaped bud is tucked up close to the body of its possessor, seemingly sheltering shyly under a bristly overhanging ledge. Its lack of adornment or detail implies a nondescript occupation. But its function far exceeds its elementary form, for this is one of two vital balancing organs that allow flies to fly so well.

Almost all flying insects have four wings: two on the middle section of the thorax and two on the rear section. In dragonflies and damselflies the wings are similarly shaped. In butterflies and moths the hind wings are slightly smaller and more rounded than the front pair, but nevertheless all four are obvious. In bees and wasps the back pair are much smaller than the front wings, but there are still four. In beetles, the front wings are hardened into tough wing cases, but there are still four. In flies, however, there are only two wings, hence the scientific name of the order, Diptera, meaning two-winged. The hind wings have evolved into vestigial stubs, called halteres, and they are used by the fly to maintain its balance when in the air.

Often likened to skittles or clubs, haltere shapes are characterised by a broad blob on the end of a longer or shorter stalk, the precise shape depending on the species. When the fly is airborne, the halteres oscillate in time with the wings. Their small size and greater density, relative to the large flight wings, gives them a gyroscopic quality. Unlike the flight wings, which have both up and down and forward and back motions, the halteres hinge freely only in the vertical plane. Sensory hairs around the base of the stem detect minute differences in the fly's body position, allowing it to compensate for pitch, roll and yaw, even in strong wind currents, giving the fly its mesmerising aerial dexterity.

RIGHT
The surface of a moth's antennae is crowded with a vast array of sensing structures to detect chemicals in the air.

LEFT
Male moth antennae are often feathered, which greatly increases the surface area, and therefore also the number of sensory organs used to detect female scents.

RIGHT
The hind wings of flies are reduced to tiny buds called halteres, which are used as gyroscopic balancing organs when the insect is in flight.

LEFT
Hoverflies have become supreme aeronauts, hovering stock-still in a shaft of sunlight, or over a flower.

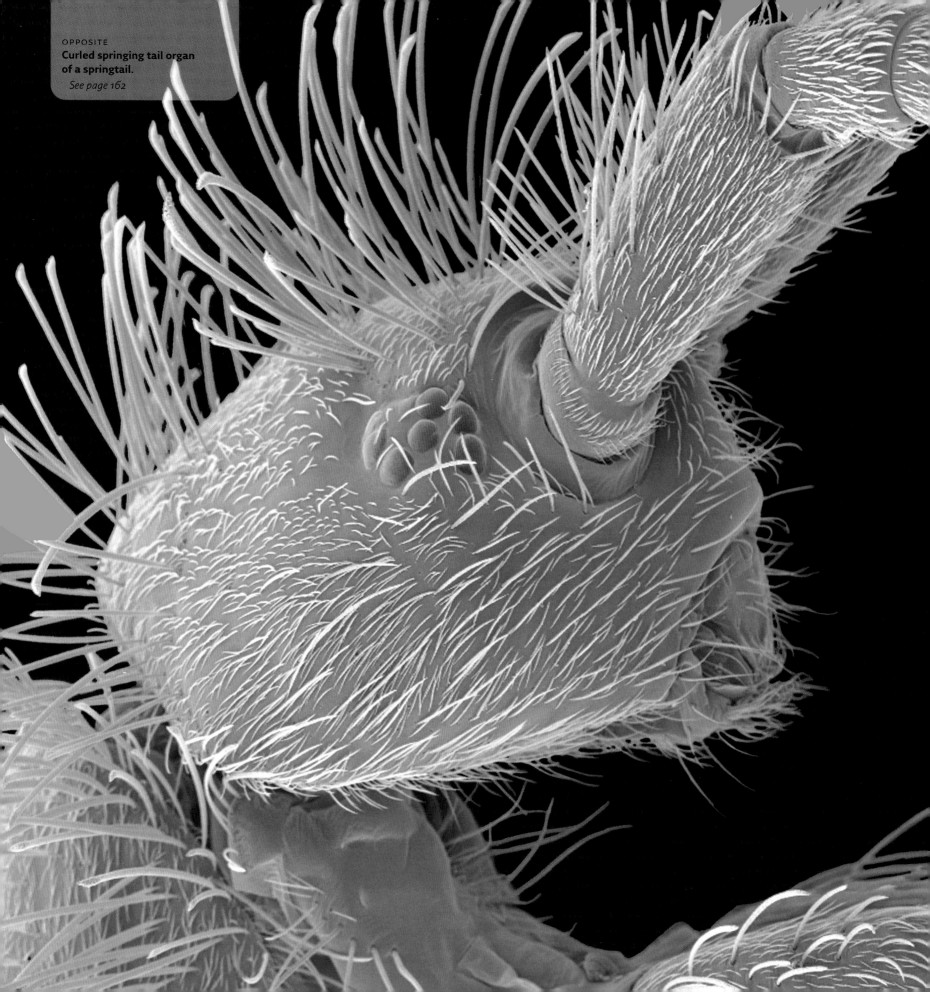

OPPOSITE
**Curled springing tail organ
of a springtail.**
See page 162

The spinnerets at the tip of the spider's abdomen exude silk from hundreds of tiny holes called spigots.

See page 162

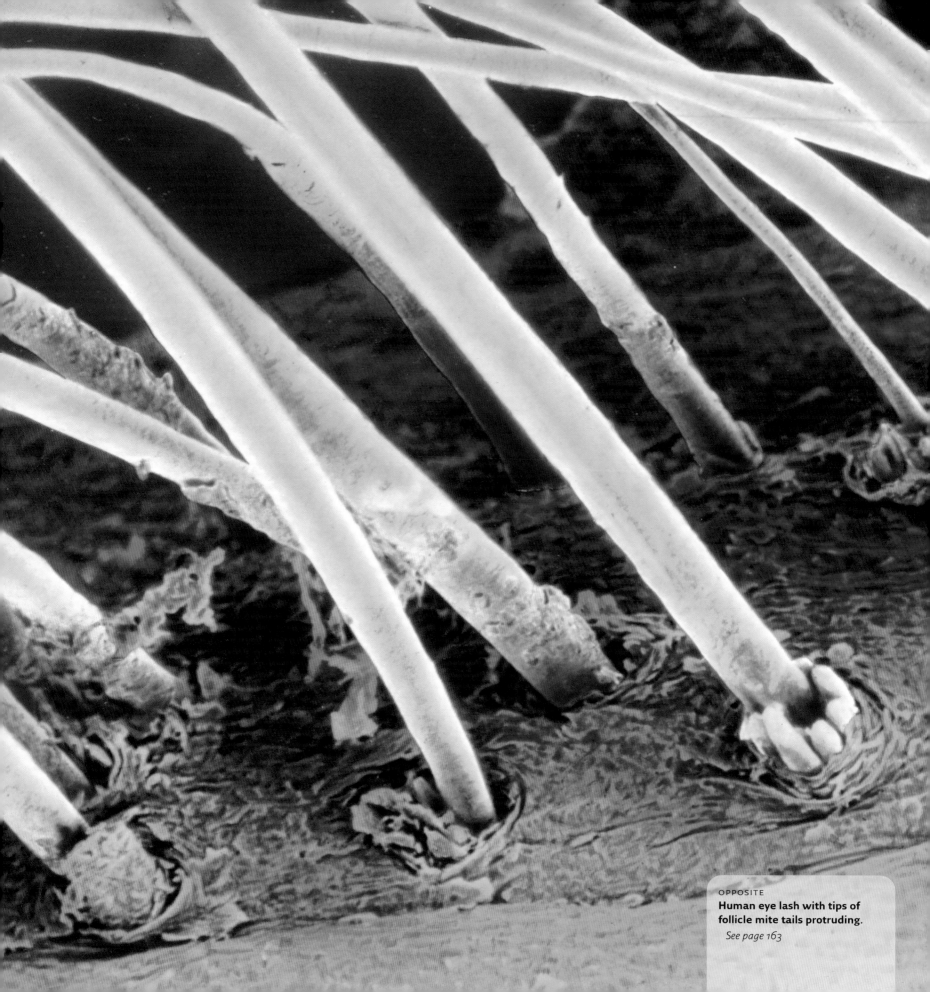

OPPOSITE
Human eye lash with tips of follicle mite tails protruding.
See page 163

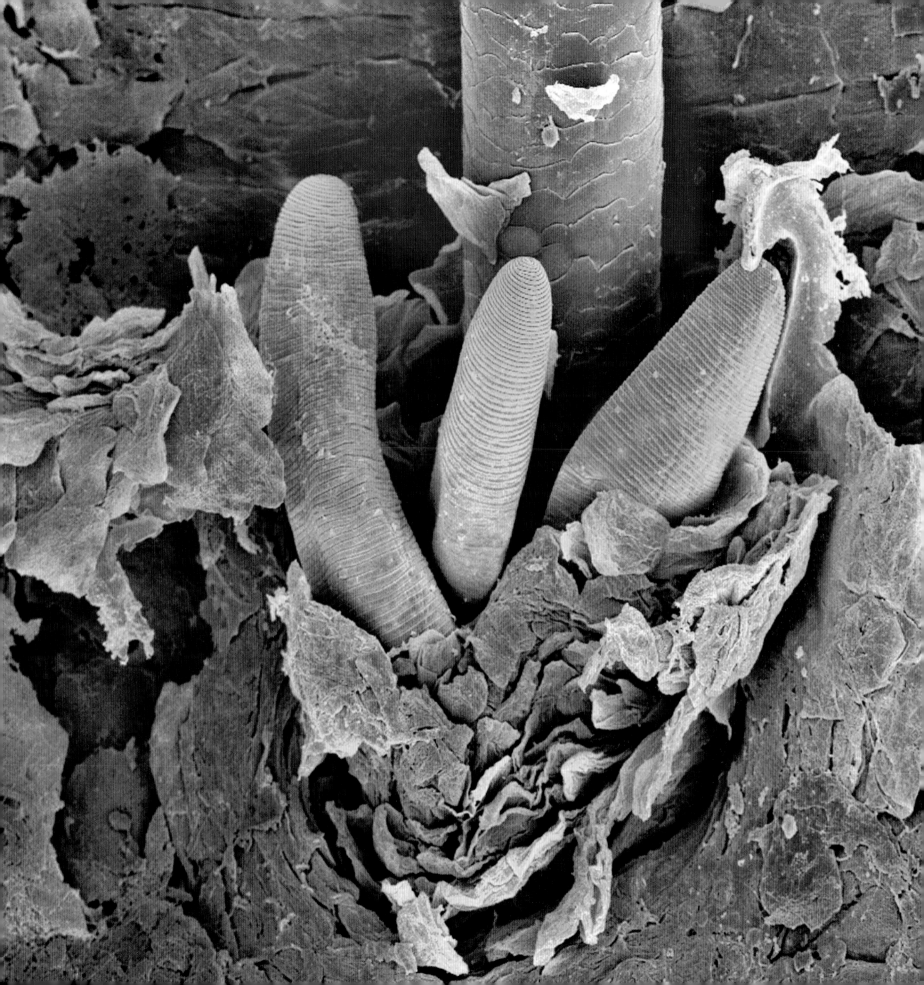

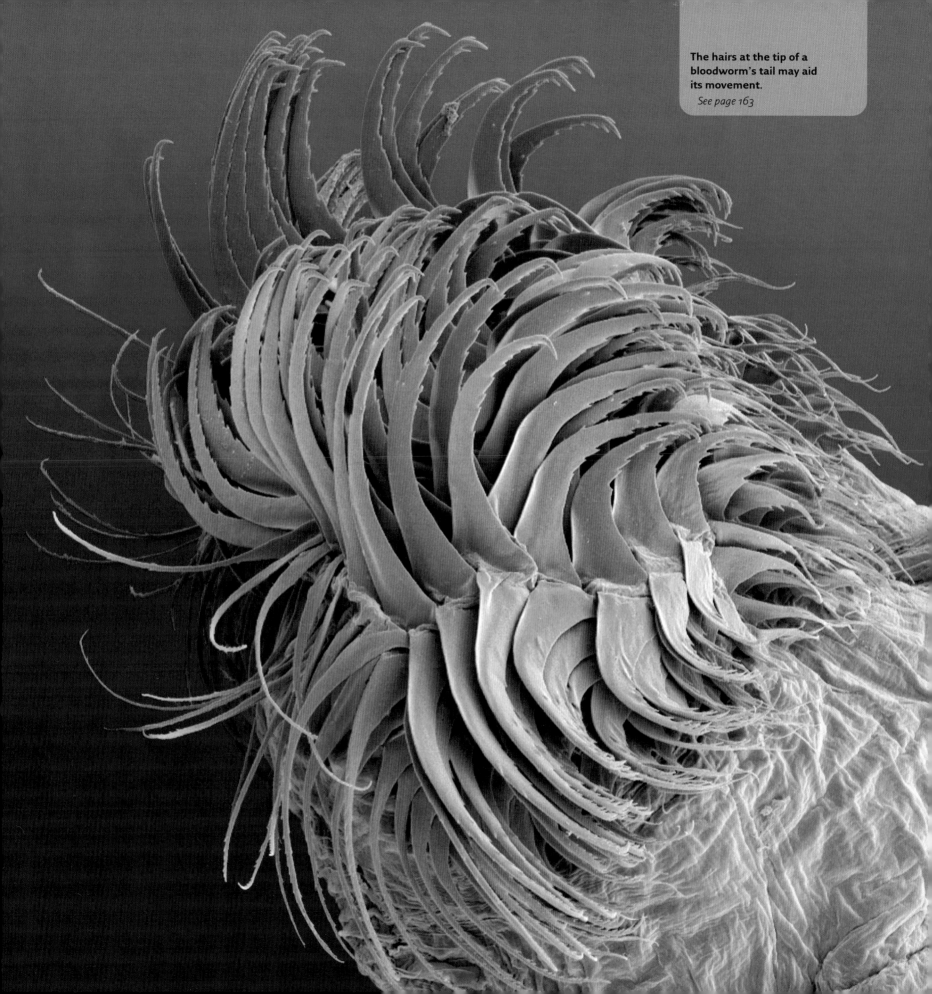

The hairs at the tip of a
bloodworm's tail may aid
its movement.
See page 163

This is not a soft curled feather boa to drape delicately around the neck. The long narrow feathers are hair-covered bristles, and the sweeping twist is a tautly coiled spring to decorate the tail of a minute animal.

The springtail is well named, and the spring in its tail is the perfect means of escape for a tiny wingless arthropod. Although springtails (scientific name Collembola) have the usual insect complement of six legs, they are now regarded as a more primitive group of creatures that have evolved separately. Like insects they have eyes and antennae, but they have no vestiges of wings. Underneath the fourth abdominal segment is a forked 'tail' (called the furcula), which has evolved from two appendages fused together. In its normal life, the furcula is tucked tightly underneath the springtail's body, held in place by a small nook in the third abdominal segment. But if the springtail is attacked or disturbed, it suddenly springs open the furcula, flicking it against the ground and flinging itself, spiralling, through the air. Although only a millimetre or two long, springtails can 'fly' 30–50 cm in a single bound, and are quickly out of danger from the confused predator that is left behind.

The evolution of flight is one of the most important reasons that insects are so numerous and so diverse in the world. But not having wings has not prevented springtails achieving a similar success. They are widely thought to be the most numerous animals on the planet, and in tropical forests densities of 200,000 per square metre have been recorded. There are over 6,000 known species, but springtails have been poorly studied by entomologists and this number is rapidly increasing as more are discovered. It has been suggested that there may well be 50,000 species out there still waiting to be found.

Resembling short hairy gloved fingers, or the gripping chuck of an electric drill, these stubby protrusions certainly look as if they work together for some particular function. They do, for they are spinnerets, the silk-spinning organs near the tip of a spider's abdomen.

The exact number and arrangement of spinnerets varies between species, but the most typical arrangement is of six, in three pairs. Although there seem to be five large organs, only four of them are spinnerets, those with a distinct seg-mented tip. Two more spinnerets are almost hidden in the centre of the group.

Silk is the secret of spiders' success, and is used for many different purposes. These include snaring prey, trip wires, abseiling and safety lines, wrapping prey, wrapping eggs. And to supply these needs, spiders have evolved at least seven different kinds of silk.

Silks are protein secretions based on a series of biochemical molecules called spidroins. Different combinations of these spidroins produce silks with different physical properties, although they are all based on a similar production process that creates an inner core and an outer sheath. The secretions are produced in special silk glands inside the spider's abdomen and flow like a liquid through progressively narrower tubules. As the silk nears the tiny holes at the tip the spinnerets (spigots), cells lining the flow tubes extract water, making it more viscous. Finally, a mild acid bath and the pulling tension of the spider's feet solidifies the silk into an elastic strand.

A silk strand is typically only 1–4 μm thick, but it has incredible tensile strength. A typical orb web may contain 20 metres of silk, combined at over 1,000 junctions and weighing only half a milligram, yet it can support a spider 4,000 times as heavy.

RIGHT
The feathery hairs on the forked tail (or furcula) of a springtail give purchase when this tiny animal flicks itself into the air, ejecting itself away from danger.

LEFT
Springtails are simple, wingless, six-legged creatures closely related to insects.

RIGHT
The spinnerets at the tip of the spider's abdomen each exude silk from hundreds of tiny holes, called spigots, before they are pulled together to form a single tough strand.

LEFT
Spiders spin flocculent (woolly) silk to disguise their web and to wrap their struggling prey.

These three sprouting buds look ready to grow, like spring asparagus tips. A fourth stem looks as if it has already shot up. But while the slim scaly stalk is genuinely growing outwards and upwards, the stout ribbed trio are pushing downwards and inwards. The stalk is a human hair, but the three 'buds' are the tails of follicle mites, tiny animals burrowing into the skin.

Disturbing as this image might at first appear, follicle mites are mostly harmless, and the majority of people who have them do not even know they are there. So common are they, in fact, but so symptomless, that estimates of their infestation rates are difficult to assess. What is clear is that we accumulate more mites as we get older, and rates of over 95 per cent are quoted for 'older' people (over the age of 65).

Follicle mites are between 0.1 and 0.4 mm long, making them among the smallest of all known arthropods.

They have a pale, semi-translucent, unsegmented cigar-shaped body and eight short stumpy legs near the head. They live almost all their lives fully or half-buried in the hair follicle, where they feed on skin cells and the oily secretions produced by the hair root. Their diet is so perfect for their growth and development that a follicle mite has no anus and excretes no waste material.

The mites occur in the hair follicles of the face, mainly around the nose, ears and eyes, and are sometimes referred to as eyelash mites. They mate, lay eggs and die in the follicle. A female mite lays about 25 eggs, which hatch after three or four days into six-legged nymphs. They are mature in three weeks, by which time they will have grown their final pair of legs. If they ever leave the follicle (which they must do to spread to a new host human during face-to-face contact), they can crawl at about 14 cm an hour. Life is slow but sure for the follicle mite.

This fanciful mop hairdo is not at the head end of an animal, it is at the tail. The curved and flattened hairs form a fringe at the very tip of the body of a larva of a non-biting midge, a *Chironomus* species. The adults closely resemble mosquitoes and biting midges but are not bloodsuckers.

The larvae, however, are often called bloodworms. They take this name from the fact that they are bright red, not with blood that they have eaten, but from the same red pigment found in vertebrate blood – haemoglobin.

Haemoglobins are very complex macro-molecules. Each is made up of smaller protein chains, twisted into globular spiral formations (globins) and attached to haem units, each of which contains a single atom of iron. It is the iron which gives the molecule its red colour and its capacity for transporting oxygen and carbon dioxide.

In vertebrates this has allowed the development of large body size, since oxygen absorbed through the lungs (or gills) can be carried to other parts of the body in the blood, and carbon dioxide can be removed.

The haemoglobin in the bloodworm is significantly different to those found in vertebrates, but it is also a complex of protein chains and iron-haem groups. Instead of using haemoglobin to transport oxygen large distances through its body, the larva uses its efficient oxygen-binding qualities to extract enough to live on, in poorly oxygenated (often polluted) water where other organisms cannot survive.

Unlike mosquito larvae, which float and breathe air through tail tubes, blood worms are bottom-dwellers. They swim by vigorously flexing their long bodies in curves, pushing a vortex of water backwards to the tail. The tufts of hairs on the tail may serve to control the direction and flow of the vortex as they swim.

RIGHT
The blunt rounded tails of three follicle mites are all that is visible as these strange animals burrow into the skin. They do their host no harm, so are called 'commensal' rather than parasitic organisms.

LEFT
Eyelashes are large stiff hairs, the purpose of which is to prevent foreign bodies getting into the eye.

RIGHT
The hairs at the tip of a bloodworm's tail may aid its movement through the water or as it crawls over the bottom of the pond, ditch, flooded tree hole or bromeliad in which it can survive.

LEFT
The aquatic bloodworm gets its name from the red of the haemoglobin in its body.

OPPOSITE
A damselfly penis.
See page 172

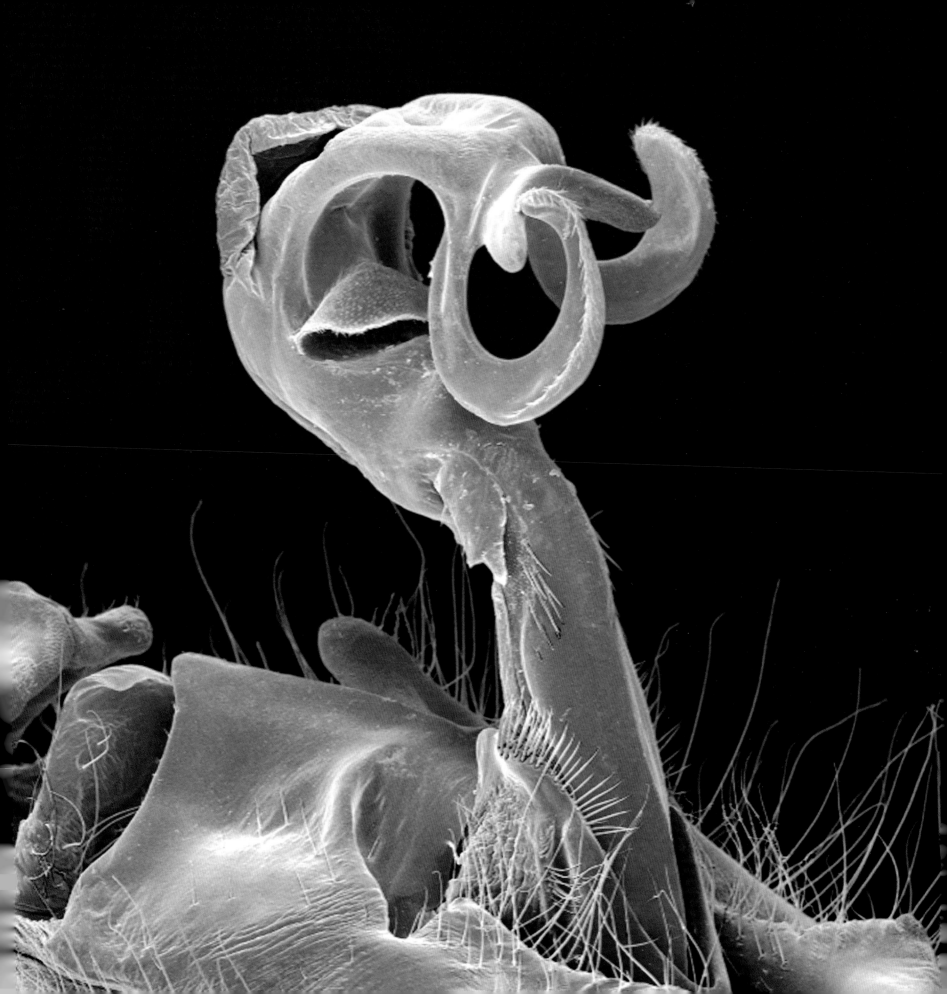

OPPOSITE
The ovipositor of a fruit fly squeezes out one of the 10-20 eggs maturing at any one time in the fly's ovaries.
See page 172

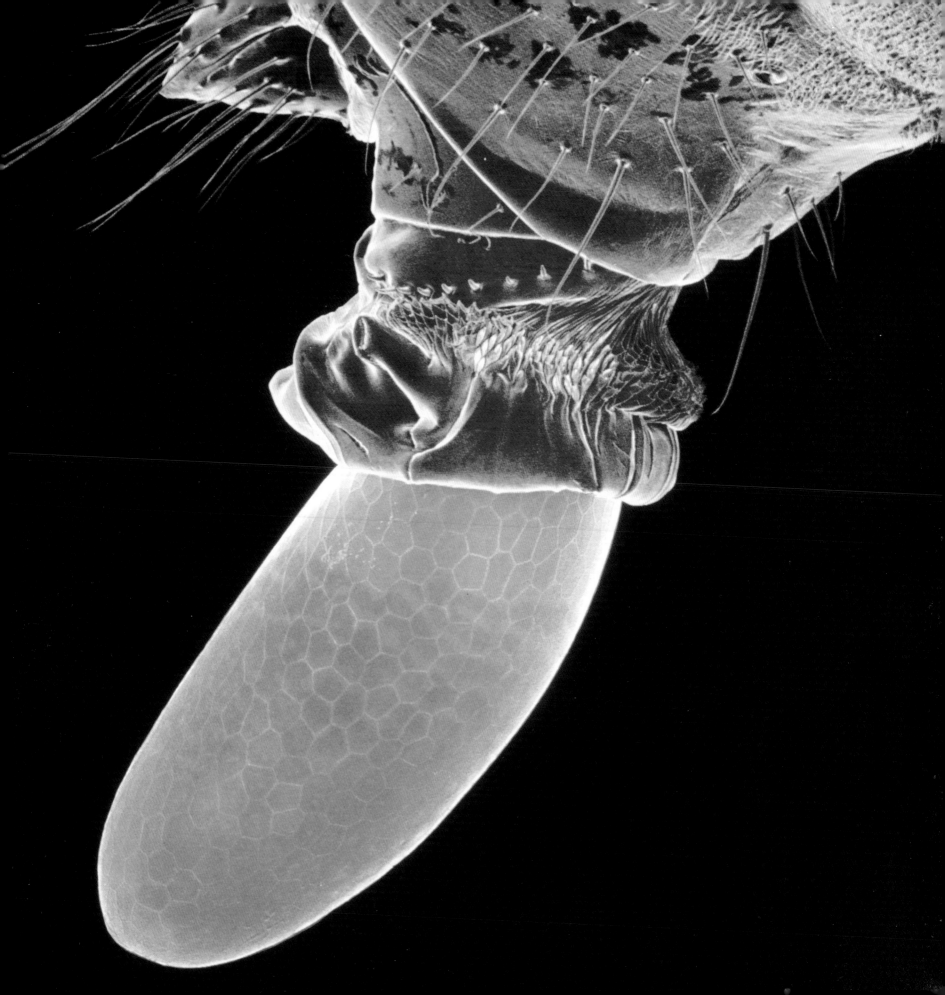

A caterpillar emerges from the egg of a large white butterfly.
See page 173

OPPOSITE
**The endophallus of a
honeybee drone.**
See page 173

This sort of construction might look more at home on a mobile phone mast or radar installation. With its curious twisted tails and contorted stem rising up through folded armour plates, a damselfly penis looks more like a weapon of war than an instrument of procreation.

The mating systems of insects are complex, and often bizarre, but they all rely on internal insemination of the female's eggs by sperm injected by the male. Technically, the portion of the male insect's sexual organ inserted into the female is called the aedeagus, and many are strangely and intricately shaped, showing great differences even between the closest of related species. It has been proposed that these exaggerated aedeagus forms have evolved much like keys, fitting only the appropriate 'locks' in females of the same species. This is to avoid cross-breeding, which is likely to be unsuccessful because the sperm/egg

product is usually infertile.

In dragonflies and damselflies (and other insects, as it turns out), injection of male sperm is only one function of the penis. The other is to remove any sperm stored by the female from previous matings with different males. This confers a huge advantage on the final male to mate, because he has a much greater chance of siring all of the female's offspring. This has led to the evolution of mate-guarding behaviour throughout dragonflies, where the male either clings relentlessly to the female while she lays her eggs, or hovers nearby, aggressively chasing off possible interloping suitors. In damselflies, it is common for the male to remain attached to the female, gripping her neck with claspers at the end of his tail, while they both submerge completely beneath the water to lay eggs in the stems of aquatic plants.

The wrinkled, textured and hairy orifice contrasts starkly with the smooth sleek lozenge emerging from within. The pale filigree tracing a ghostly mesh of hexagonal lines across the glassy surface adds to the disparity of the image. Strange though it may appear at first sight, here is caught an act rarely observed yet so common and mundane – the laying of an egg.

The egg, one of 10 to 20 maturing at any one time in the ovaries, is emerging from the abdomen of a fruit fly. Fruit flies, especially *Drosophila melanogaster*, have become important and well-studied test organisms, particularly in the science of genetics and embryology, and much of our understanding of how animals work is extrapolated from these organisms, which are easily and quickly reared and easily manipulated.

Simple though the egg structure appears, it is more than just a simple biological entity containing the genetic material of the DNA. Even before the

larval embryo starts to develop (even before the egg is fertilized by a sperm), the egg already has a front, back, up, down, left and right determined, which will translate into these axes in the adult insect when it finally emerges after successful metamorphosis, many hundreds of thousands of cell divisions later in life.

The mechanisms determining body form and orientation are still being examined by scientists, but by using a relatively simple animal, such as *Drosophila*, an understanding of some of the basics is emerging. Before the fruit fly egg is fertilized, it is composed of 16 superficially identical cells. One of these goes on to become the oocyte (the female egg cell fertilized by a male sperm) while the other 15 become 'nurse' cells. As well as providing nutrients for the developing oocyte, these lay down a gradient of protein and genetic markers that influence the final body plan of the future fly.

RIGHT
The aedeagus (penis) of a male damselfly not only inseminates the female, its twists and curves also scoop out any competing sperm from previous mates.

LEFT
Although smaller and daintier than dragonflies, damselflies have similar aquatic nymph stages and are aerial predators of small flying insects as adults.

RIGHT
The flexible ovipositor of a fruit fly contorts and stretches to squeeze out one of the eggs maturing in the fly's ovaries.

LEFT
Its simple life history and fast generation time have allowed the fruit fly, *Drosophila melanogaster*, to become one of the major tools of biology, especially in the study of genetics.

These delicate fluted and faceted semi-transparent domes, like high-tech glasshouses, hold a secret: in each of them a monster is waiting to emerge, and one has already started to break free. At barely 2 mm long, the caterpillar is hardly monstrous yet, but it will eventually grow into a major agricultural and garden pest, the large white butterfly, and by those who grow cabbages it is rightly regarded as a monster.

The ribbed barrel-shaped eggs, bright yellow in life, were laid about a fortnight ago, in a tight batch of 20–100 on the underside of a cabbage leaf. The whole batch will hatch at more or less the same time, and it is not long before a squadron of caterpillars, feeding side by side, is on the march. At first they graze the lower layer of plant cells, leaving translucent windows in the leaf; then, when their mandibles are big enough to manage the leaf thickness, they start shredding the plants down to a skeleton of veins and stalks.

The gregarious behaviour of the caterpillars is a direct result of their diet. Despite the palatability of cabbage to humans, the uncooked leaves contain high concentrations of toxins called glucosinolates, which are poisonous to many animals. But the large white caterpillar has evolved to cope with these noxious chemicals. Not only is it immune to the poisons, it actively harvests them, mainly one called sinigrin, and stores it in its body. This makes the caterpillar distasteful to any bird or animal that might try to eat it.

The caterpillar advertises its dangerousness by adopting warning colours: bright yellow covered with prominent black speckles. It does not need to hide from predators, but instead gains extra protection by feeding in an even more warningly coloured group.

There is nothing funny about this jester's cap, with its twisted conical leathery horns and brushes of feathery tufts. This is the most important thing a male honeybee possesses – its penis.

A honeybee colony runs its everyday life without males. A single fertile female (the queen) spends her time laying eggs while many tens of thousands of infertile females (workers) forage for pollen and nectar, feed the brood of grubs, construct the wax combs, clean and clear the nest of debris, and protect the colony from honey thieves. In their domesticated hives, beekeepers would have this continue for as long as possible, and until they are ready they control the size of a colony to prevent its division by swarming, which happens when the bee numbers become too great.

At some point either the number of workers or the physical size of the nest breaches a threshold beyond which chemical signals from the queen are diluted too far – and the workers change their behaviour. They start to create differently shaped cells in the comb. Some, with broader openings, stimulate the queen to deliver a male egg, while others are larger and are destined to produce new queens as the workers preferentially feed the grubs inside them the high-protein secretion sometimes called royal jelly.

The first adult queen to hatch seeks out and kills all the other queen pupae by stinging them, then leaves the nest to mate, pursued by ardent drones. Mating takes place on the wing, and after ejaculation the male departs. But his endophallus, the section of the penis inserted into the female during copulation, is broken off. The drone which suffered this catastrophic castration will soon die. The next drone to mate is able to scoop out the previous male's broken endophallus and deposit his own sperm. After mating several times, the queen has enough sperm stored in her spermatheca, a special reservoir in her abdomen, to last her lifetime.

RIGHT
Butterfly eggs are beautifully sculpted, with seams and pleats giving corrugated strength to their delicate shells.

LEFT
The large white caterpillars feed gregariously, gaining safety in numbers and advertising their foul taste by warning colours.

RIGHT
The honeybee penis snaps off inside the queen during mating, but the next male to copulate with her can scoop out the remains and add to the queen's sperm store.

LEFT
Apart from mating with the queen, the honeybee males (drones) serve no purpose in the hive.

A stoma on the underside of a tobacco leaf, surrounded by two guard cells.
See page 182

OPPOSITE
A spiracle, a breathing hole along the side of a caterpillar.
See page 182

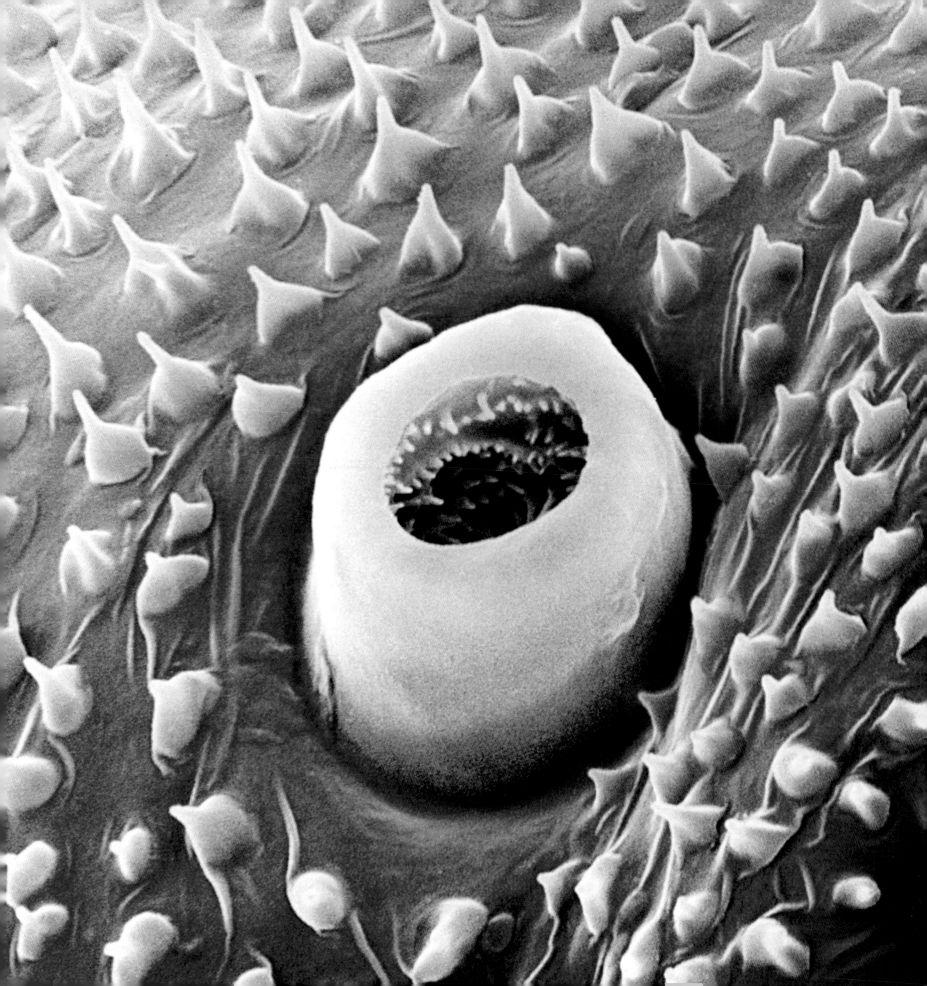

OPPOSITE
Gills along the body of the aquatic nymph of an alderfly.
See page 183

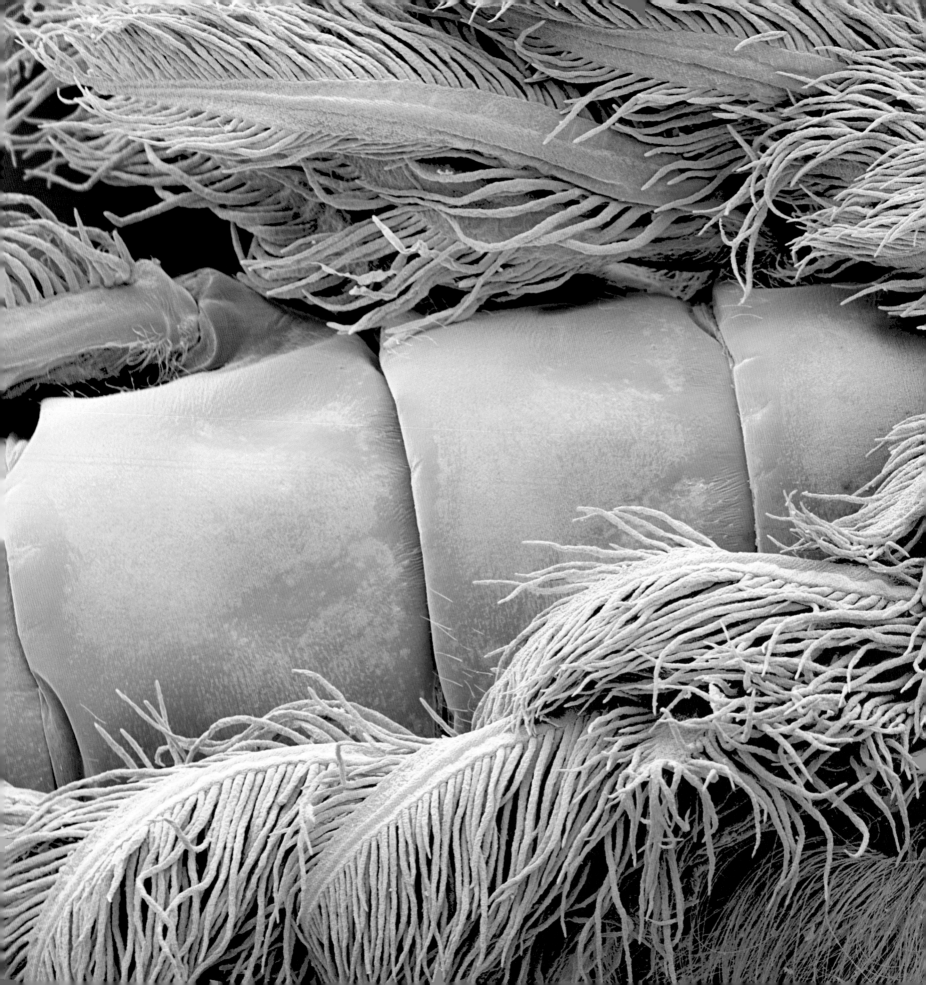

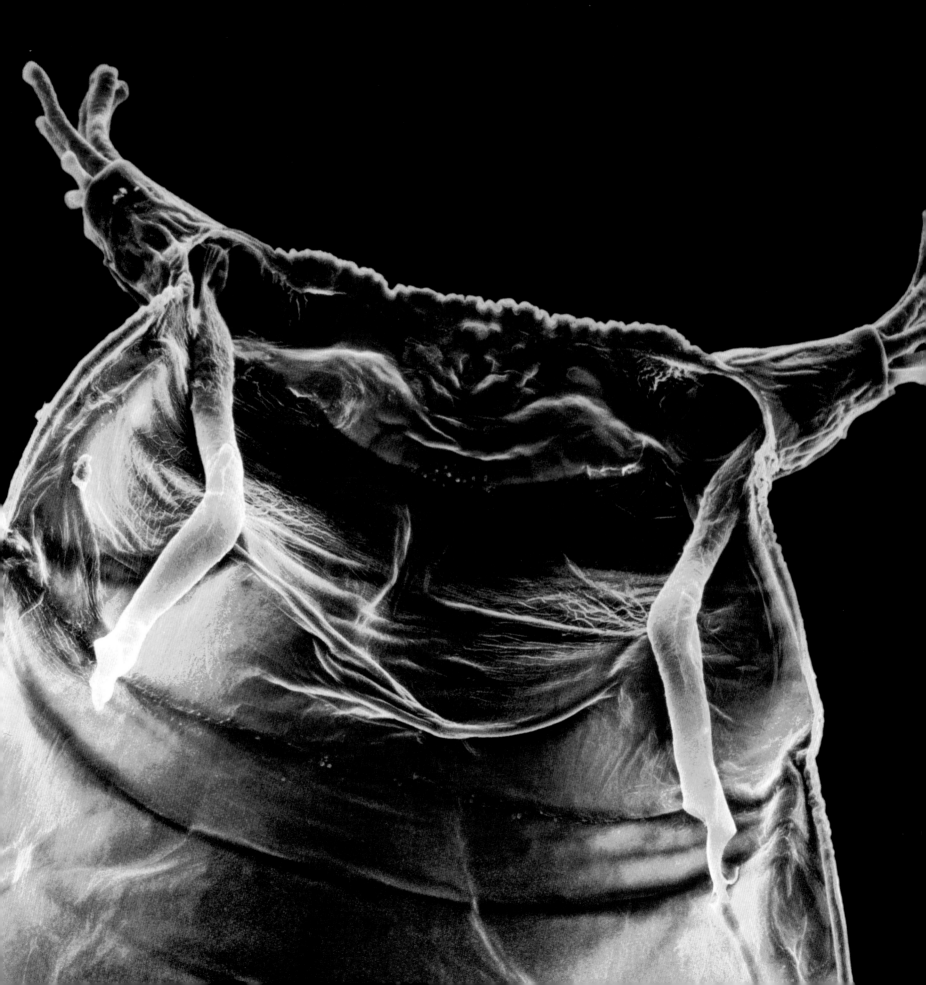

The breathing tubes (spiracular papillae) at the tip of the puparium allow the developing fly to breathe.

See page 183

Whatever they are attached to, these are lips, pert and pouting. And, as with all lips, the space between them is precisely aligned to control the passage of air. But no sound emanates from these, since they are the lips around a breathing pore on a tobacco leaf.

Plants regulate water loss and gas exchange through these pores, or stomata. In dicotyledons (most broad-leaved flowering plants and trees) the majority are on the underside of the leaves; in monocotyledons (grasses, irises, lilies etc) they appear equally on both surfaces.

When opened, each stoma can release water vapour through evaporation, causing a negative water pressure inside the plant that helps suck up more water (and mineral nutrients) through the roots. The stoma also allows the plant to take in carbon dioxide from the atmosphere. Using the energy in sunlight, it chemically combines this gas with water, in the complex process of photosynthesis, to produce simple

carbohydrate sugars. Oxygen, created as a by-product, is released.

Each stoma is bounded by two guard cells, the lips around this mouth. Stomatal opening is controlled by the flow of water molecules into the slightly curved guard cells, which become bloated and bulge outwards, creating an oval hole. When the water molecules are released from the guard cells, they relax and deflate so that the gap between them closes.

The precise mechanism of water flow is not fully understood. It was formerly supposed that guard cells created sugars by photosynthesis during daylight, since they contain chlorophyll (the green pigment that harnesses light) whereas other leaf surface cells do not, and that as the sugar concentration increased it drew in water molecules from surrounding cells by osmosis. But recent studies have suggested that this mechanism is not sufficiently quick or powerful to account for stomatal opening and closing.

The gelatinous doughnut ring surrounded by tiny peaks of pink icing would not look out of place on a child's birthday cake. This cake decoration is not, however, merely placed on the surface; it is the outer ring protecting a hollow tube that descends far below the surface – a breathing hole for a caterpillar.

All insects have breathing holes, called spiracles, along the sides of their bodies. Moth caterpillars have nine pairs – eight on the abdominal segments, and one on the front thorax segment (the developing wing buds replace the spiracles on the middle and hind thorax sections). Each spiracle is the entrance to a tube that branches again and again as it penetrates the segment. The tubes, called tracheae, allow oxygen in the air to reach into the interior of the caterpillar to be used in the insect's metabolism, and then allow the carbon dioxide that is produced to escape.

The muscular movement of the caterpillar, as it walks about on its food-plant, helps pump air to and fro along the tracheae. In addition a rudimentary pump in the head capsule sucks haemolymph (the insect equivalent of blood) along the only true blood vessel down the caterpillar's back and pumps it out into the body cavity. As the haemolymph flows passively back towards the caterpillar's tail it absorbs and transports some of the oxygen, ensuring that no part of the animal suffocates.

Insects, including caterpillars, can control the opening and closing of the spiracles by a valve just beneath the surface (not quite visible here). In caterpillars the valves mainly close to control water loss during hot weather, but also shut to stop the caterpillars drowning if they are inundated by rising water. This feature allows many insects to live in the tidal zone or in marshy places that regularly flood.

RIGHT
On the underside of the leaf, two lip-like guard cells expand or relax to open or close a gas-exchange pore.

LEFT
Tobacco, like all green plants, uses the sun's energy to drive photosynthesis, creating carbohydrates from water and carbon dioxide.

RIGHT
The raised ring is the opening of a breathing pore, called a spiracle, on the caterpillar of a garden tiger moth, one of nine down each side of the animal's body.

LEFT
Sometimes called a woolly bear, the furry bristles of the tiger moth caterpillar offer some protection against predators.

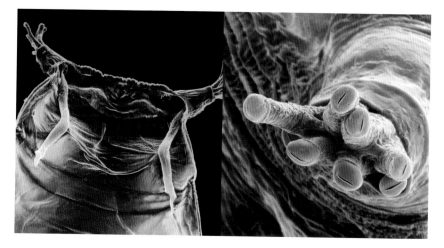

These fancy feathers do not serve to keep a bird warm, or decorate its body – nor do they flutter lightly in the wind, for this is an aquatic animal, and the highly branched fronds are its breathing apparatus, its gills.

Alderflies form a small group of primitive insects belonging to the order Megaloptera. Adults have four large slow-flapping wings, and are similar to ant-lions, lacewings and dragonflies. They always occur near water because their nymphs (larvae) are aquatic, living in freshwater ponds, lakes and streams. Fresh water is often alive with insects, but many, like mosquito larvae, water beetles and boatmen, have to visit the surface to take in air to breathe. Truly aquatic animals, like the alderfly larva, have gills to absorb oxygen dissolved in the water.

Air contains about 200,000 ppm (parts per million) oxygen, that's 20 per cent. By contrast, water fully saturated with dissolved oxygen contains only 15 ppm. But it is not this difference that makes breathing under water so much more difficult for insects, it is because water is so much denser than air. The comparative heaviness of water makes it impossible to pump it in and out fast enough to gather oxygen.

Gilled insects overcome this problem by having hugely branched and developed external gills, offering a much greater surface area to the water flowing past outside, through which oxygen molecules can be absorbed.

Having a pair of gills on each body segment is an echo of the evolutionary descent of the gills from the tracheae, the internal breathing tubes found in terrestrial larval and adult insects. This is confirmed by the fact that the gills have gas-filled transport tubules inside them. Passive movement of the oxygen through the gas is several hundred thousand times faster than through liquid.

These finger-like extensions seem to reach, outstretched, directly towards the viewer. Or perhaps they are some other sense organ, like the eyed tentacles of a snail, their slit pupils craning for a better view. At the tail end of a fruit fly pupa, these tubes are performing another function, for they are the animal's breathing tubes and the slits are the narrow holes through which oxygen is taken in and carbon dioxide expelled.

Unlike caterpillars, which have spiracles in almost every body segment, fly maggots are usually limited to breathing tubes at the tips of their tails (although in a few there are spiracles at the front of the body also). This is a good reflection of the maggots' lifestyle, which often means feeding in liquid or semi-liquid decay such as carrion, animal dung, compost, manure or stagnant water. Others are hidden under the soil or inside plants. Having a breathing tube at the tail end allows the larva to feed with the head end down. In mosquito larvae, the posterior spiracles are surrounded by hairs which close like a valve when they submerge in water, but which flap open to create a snorkel when they return to the surface. Rat-tailed hoverfly larvae live in liquid sewage and can have breathing tubes (their 'rat tails') nearly 30 cm long.

This is the pupa of a fruit fly. The larva has been feeding in rotten fruit, and although it may have moved a short distance to pupate, a stiff upright array of breathing tubes will help it survive if the putrescent fruit juices threaten to inundate it.

Strictly speaking, the hard shell, inside which the adult fly develops, is the inflated and toughened skin of the larva (the puparium); the true pupa or chrysalis forms inside. These posterior spiracle tubes now let air in to the interior of the puparium, so that the pupa can still breathe during its final metamorphosis.

RIGHT
Feathery gills extend from each body segment of an alderfly larva. The many-branched fronds allow a huge surface area through which to absorb oxygen dissolved in the water.

LEFT
Alderflies are related to lacewings; their aquatic larvae are fierce predators.

RIGHT
The breathing tubes at the tip of the puparium are called spiracular papillae. They allow it to breathe while the adult fly develops inside.

LEFT
The puparium of the fruit fly consists of the inflated and hardened skin of the maggot, inside which the true chrysalis contains the metamorphosing fly.

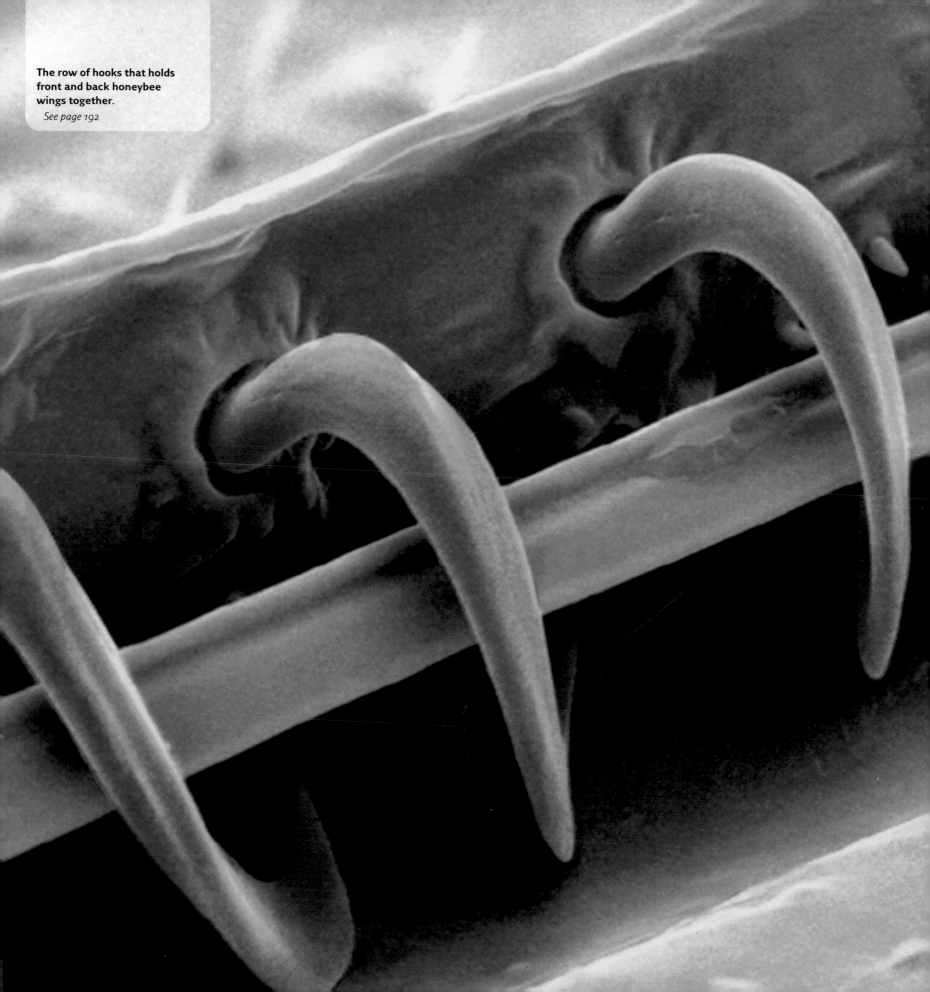

The row of hooks that holds front and back honeybee wings together.

See page 192

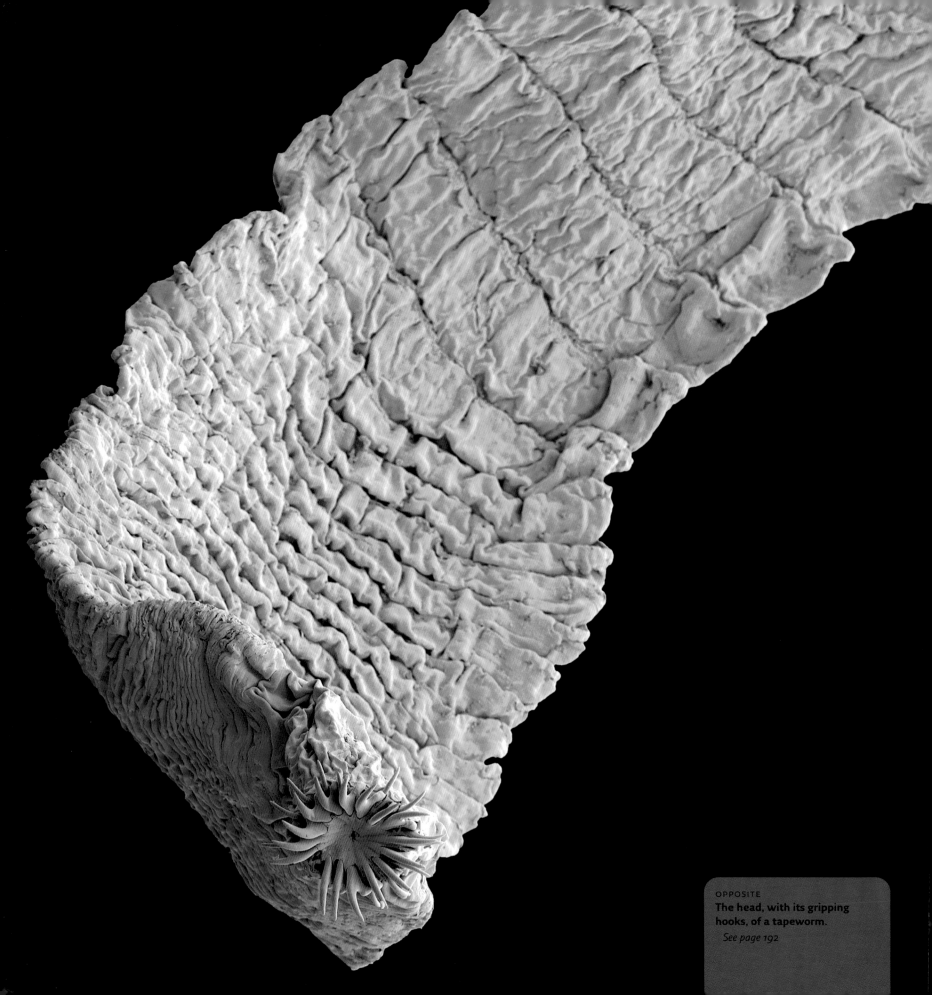

OPPOSITE
**The head, with its gripping
hooks, of a tapeworm.**
See page 192

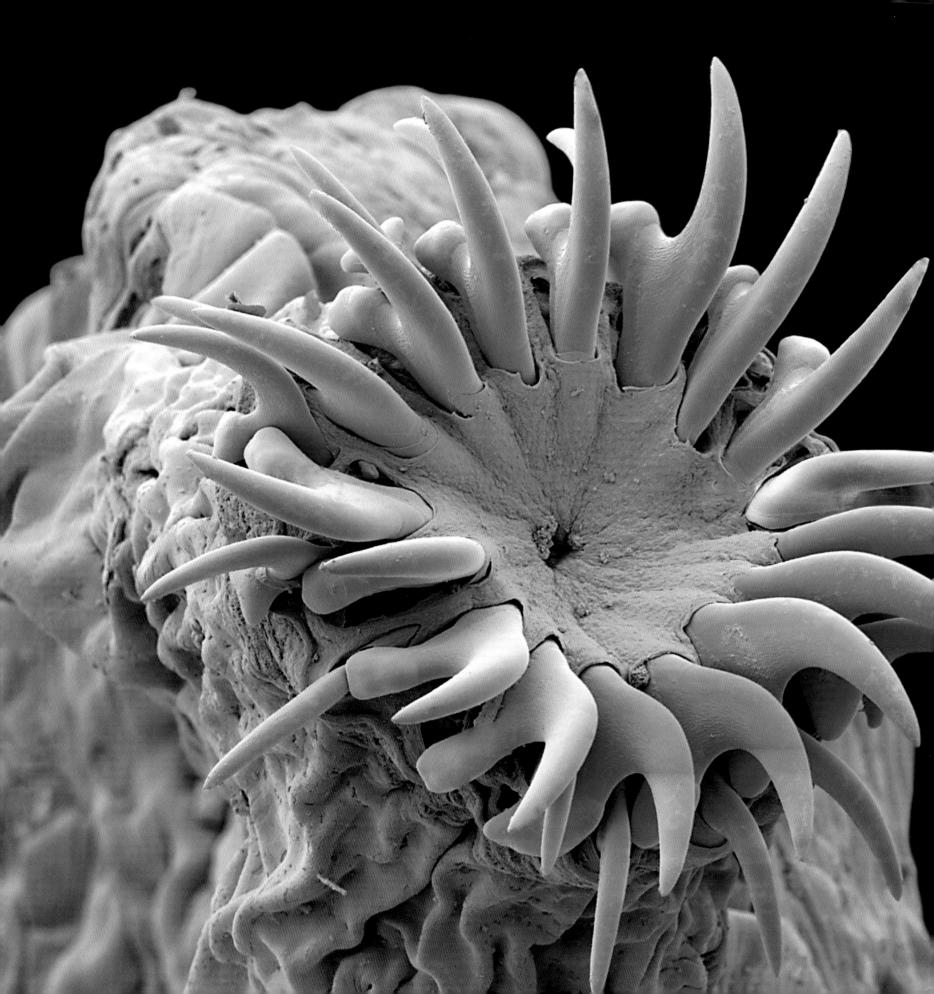

OPPOSITE
The aedeagus, the penetrative male organ, of a bed bug.
See page 193

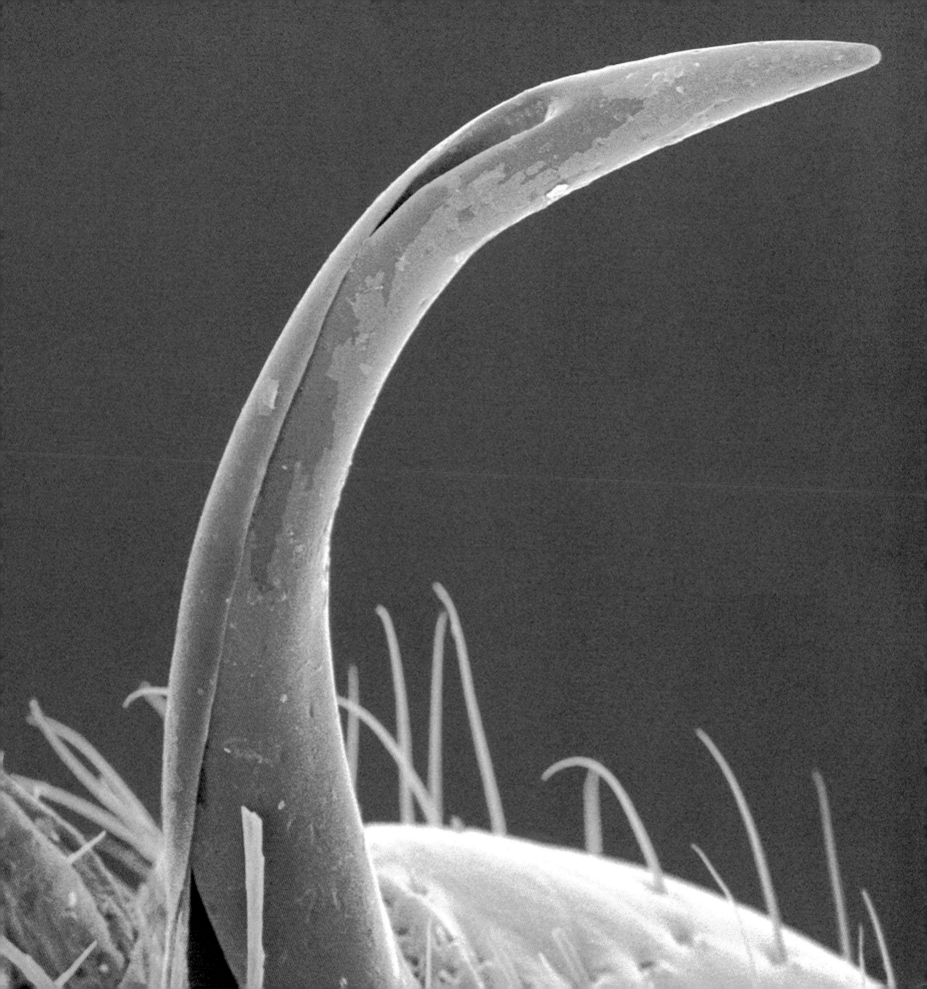

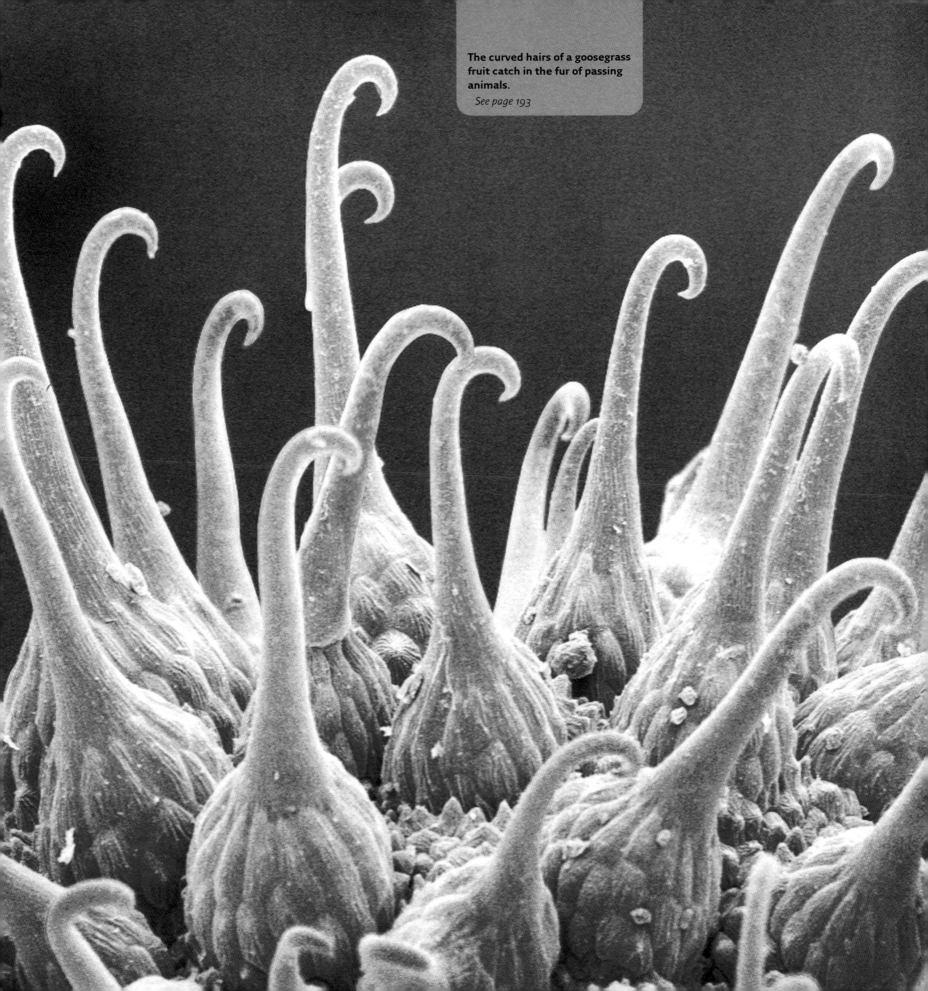

The curved hairs of a goosegrass fruit catch in the fur of passing animals.

See page 193

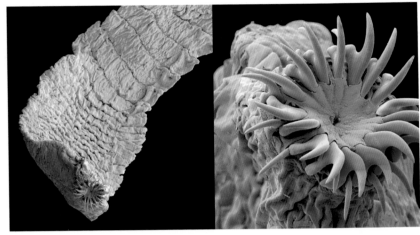

Like a series of wrought-iron hooks surrounding a steel pole, this could be the hinge of an elaborate garden gate. Or it could be the delicate spiral twist of a wire-bound note pad. Instead it is a series of gripper hooks running along the edge of a honeybee's wing.

Most flying insects have four wings, one pair sprouting from the middle segment of the thorax and one pair from the rear segment (only flies, Diptera, have two wings, the back pair being reduced to knob-like balancing organs, as shown on page 151). In more primitive groups, including dragonflies, damselflies and lacewings, each pair of wings beats up and down in a movement that is synchronized with the other pair, but independent from it.

In the numerous, hugely diverse and successful insect order Hymenoptera (bees, wasps and ants) front and back wings beat together as one organ. The front wing is much larger than the hind wing, which acts more like an additional lobe extension rather than a separate limb. The front and back wings are held together by this series of curved hooks, called hamuli, which reach out from the hind wing and grasp the reinforced trailing edge of the forewing. This increases the aerodynamic power of the wing system by having two muscular motors powering one unified flight membrane.

The mechanism of wing-coupling occurs in most other insect groups, and appears to have evolved independently. In butterflies a bulge called the humeral lobe on the hind wing grips the back edge of the front wing. In moths this is replaced by a spine called the frenulum, on the underside of the back wing, which is caught by a curved hook called the retinaculum beneath the front wing.

The double ring of stout hooks serves a vital purpose for a tapeworm: they are the anchor with which the animal attaches itself to its host's intestines, and they must be able to firmly grip the intestinal wall against the pressure of half-digested food passing through the gut. Tapeworms are so-called because they have flattened ribbon-like bodies, up to 18 metres long. All tapeworms are intestinal parasites of vertebrates, and their flattened form is an adaptation allowing them to easily absorb nutrients through their skin from the host's gut. A worm is divided into hundreds of segments called proglottids. Each proglottid contains a uterus, inside which are up to 50,000 eggs, along with male reproductive organs. As the worm matures it sheds segments from its tail, and these pass out of the anus in the faeces.

The minute eggs pass into the environment they are inadvertently eaten by a secondary host animal. This dog or fox tapeworm has rabbit and hare as its intermediate carriers. The tiny eggs are small enough to escape being chewed, and inside the stomach they react to the digestive enzymes by invading the gut wall and migrating out into the muscle tissue. Here they form small cysts, strictly called cysticerci, or sometimes bladder worms. It is by eating meat infested with cysticerci that a dog acquires a new tapeworm inside its intestines.

Humans can be infected by two species of tapeworm. *Taenia saginata* infects cattle as its secondary host, and *Taenia solium* infects pigs. Humans become infected if meat containing the cysticerci is not properly cooked. Symptoms of infection include malnutrition and intestinal blockage.

RIGHT
A row of hooks called hamuli, along a honeybee's hind wing, grip the front wing to create a combined surface when the bee is flying.

LEFT
Honeybees are strong fliers, and regularly forage several kilometres from the hive after nectar and pollen.

RIGHT
The ring of hooks around the head of a dog tapeworm anchors it in the animal's intestines while the remainder of the long thin body absorbs nutrients from the part-digested food in the gut.

LEFT
The flat body shape offers a large surface area to absorb nutrients from the host's intestinal tract.

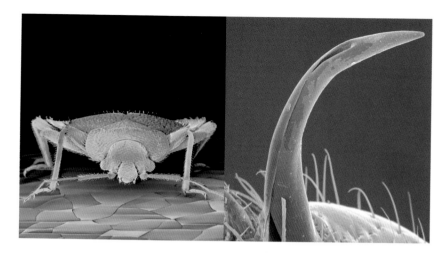

Rising like a giant sail-maker's needle, ready to receive its thread, this hook looks set to penetrate even the toughest material. Hooks are useful tools, and occur widely through the plant and animal kingdoms. They are used as tools for gripping and holding, but also as weapons for piercing and tearing. In this instance, the bedbug's penis verges on both tool and weapon, for during the process of mating he uses the curved spike to rupture the female's abdomen and inject his sperm into her body cavity.

The spike, called a paramere, is only a part of the male bedbug's genitalia. In most insects there are usually two such structures, paired to form the aedeagus – the male's penetrative organ. During bedbug copulation, the male's paramere punctures the female's abdomen, underneath, between two of her body segments. Insertion is not arbitrary, but directed always at the same spot by a deep indent in one of the female's segmental plates. This acts as a guide,

directing the paramere to a point inside her body from where the sperm can easily migrate to the ovaries.

The bedbug's mating strategy is known as traumatic insemination, and the only other organisms known to use it are an obscure group of insects called Strepsiptera. Related to beetles, they are minute parasites of other insects, mainly bees and plant bugs. Although apparently violent, this behaviour is normal and natural, and is thought to have evolved because in almost all creatures there is a conflict between the sexes. In insects, for instance, males usually seek multiple copulation to improve their chances of siring offspring, and the energy costs of high sperm production are low. On the other hand, females can be damaged or exhausted by repeated matings. Traumatic insemination is at one end of a diverse spectrum of mating strategies and counter-strategies that have evolved in the battle of the sexes.

The formidable array of crochet hooks is ready to snag any passing object. But rather than trap and hold some defenceless victim, their purpose is more subtle – to hitch a lift. The tiny hooks surround the fruit of goosegrass, and once caught in the fur of a passing animal, their aim is to hold on until they have been moved to pastures new.

Goosegrass is not a grass, but a narrow-leaved sprawling plant with small white flowers. It grows in hedgerows, verges, rough meadows and the edges of arable fields. It is a vigorous grower and quickly shades out other plants. Simply dropping its seeds onto the ground below would be tantamount to genetic suicide for the plant, because its seedlings would be shaded out by the parent.

Instead, the tiny fruits, each containing up to three seeds, are covered all over with these tiny hooks. Fruits and seed cases covered in hooks and sticky hairs are called burs, and they have evolved in many different plant groups.

Passing animals quickly pick up the fruits, which become entangled in their hair or feathers. Since humans have been wearing clothes, they too have become the unwitting dispersers of the plant. It is well known by many appropriately adhesive names including stickywilly, stickyweed, catchweed and cleavers.

The hooks are not randomly directed, however: they all more or less point in the same direction. This helps the fruit become eventually dislodged. It is all very well getting stuck on an animal for transport, but it is also important to become disentangled, so the seed can fall to the ground and germinate.

Dispersal of offspring is important in maintaining a good population. It increases the geographic range of the organism, helps with the genetic mix by preventing local in-breeding, and avoids extinction by establishing outlying (back-up) colonies. Goosegrass has achieved these goals supremely, and is now a common herb throughout the northern hemisphere.

RIGHT
The male bedbug's penis is used to puncture the female's abdomen and insert sperm directly into her body cavity.

LEFT
The bedbug occurs in human dwellings throughout the world. Its flat shape allows it to hide in cracks in beds, behind wallpaper and other tight spaces.

RIGHT
The curved hairs of a goosegrass fruit catch in the fur of passing animals (and human clothing) and the seeds are dispersed away from the strong-growing parent.

LEFT
Goosegrass is not really a grass at all, but a narrow-leaved sprawling plant related to bedstraws.

OPPOSITE
**Opening of the spore capsule
of a moss.**
See page 202

OPPOSITE
The bead-like spores of an *Aspergillus* **fruiting body are ready to be wafted into the air.**
See page 202

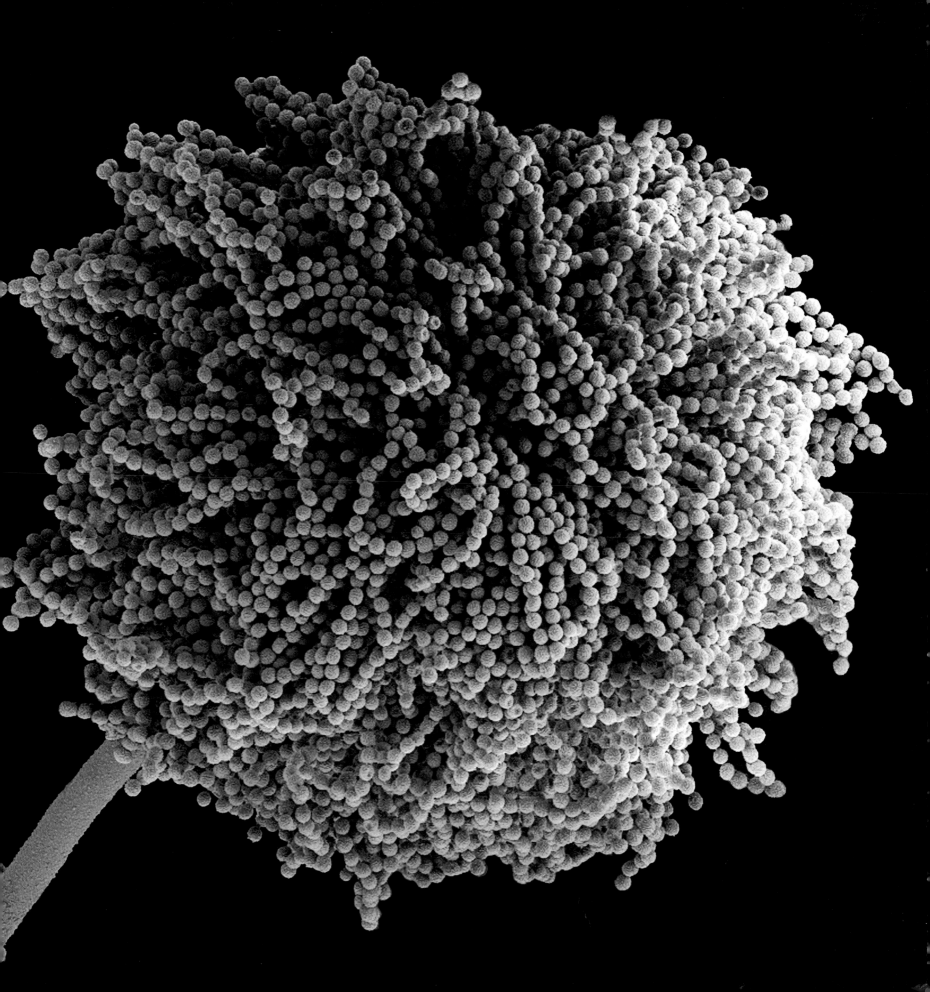

The anthers of the chameleon plant produce and release the pollen.

See page 203

OPPOSITE
**Spores on the underside
of a bracken leaf.**
See page 203

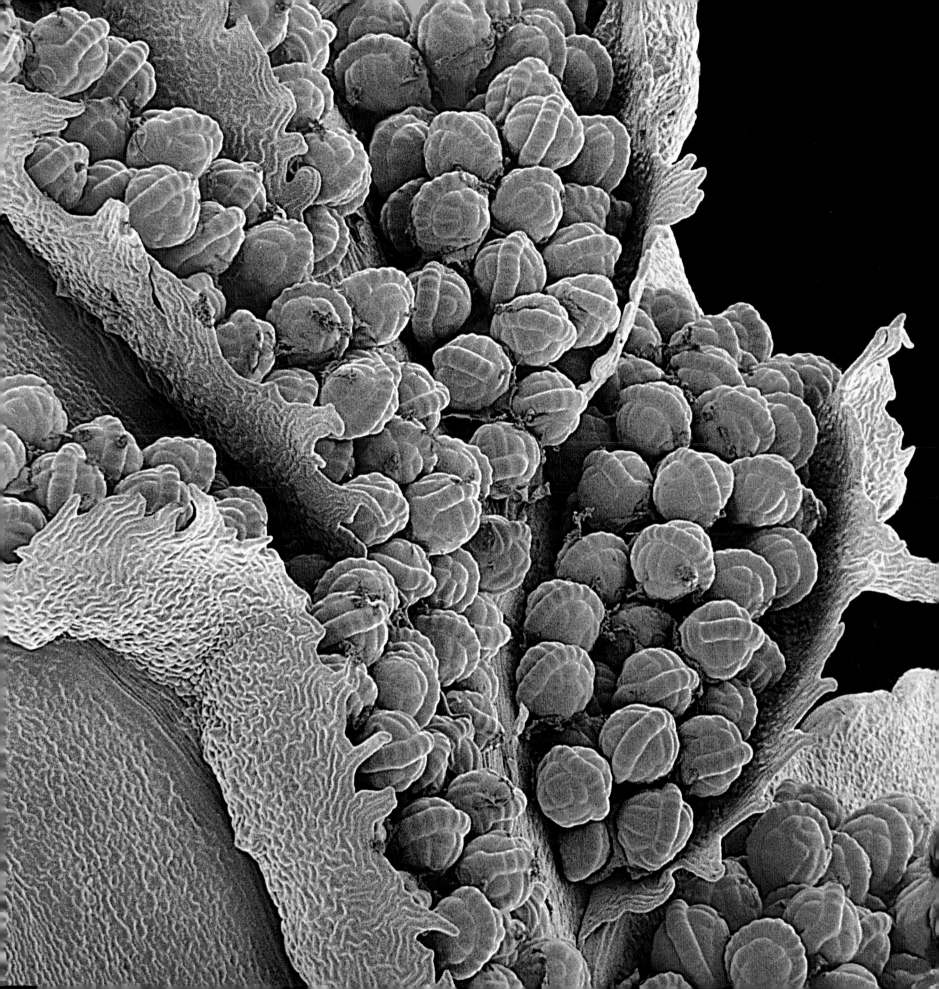

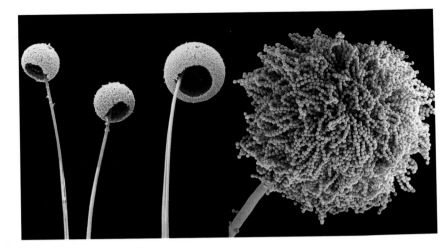

Like the air-intake nozzle of a turbo-jet engine, the toothed twisted triangular foils look as though they may control the flow of air into the machine. But they are for controlling flow out, not in – the flow of tiny spores released from the fruiting head of a moss.

Also called bryophytes, mosses are non-vascular plants: that is, they do not have the long tube-like vessels running through their roots and stems to transport food and water. Instead, they are usually little more than clumps of leaves in dense mats or tufts just a few centimetres high. The only outstanding features they do have are the sporophytes, spore-bearing 'fruits', each having a long-stemmed stalk tipped with a spore-dispensing capsule.

Unlike other green plants, mosses do not have flowers, and their spores are not the same as seeds, because they have no stored food reserves. However, spores do germinate into flat plant-like growths which then develop into the stems and leaves we recognize. Like other green plants, they use sunlight to photosynthesize nutrients. As they develop, male and female sex organs arise at the tips of stems and branches.

Mosses usually grow in wet and damp situations, because they need moisture during the process of sexual reproduction. Moss sperm, which is very similar in appearance to animal sperm, except that each spermatozoon has two tails instead of one, must swim through the thin layer of water on the leaf surfaces to fertilize the female eggs. It is a fertilized egg that now grows into the sporophyte, with the wind-blown spores, the dispersal phase of the moss's life, developing inside it.

There are about 10,000 species of moss described in the world. *Funaria hygrometrica* is common throughout the globe, particularly growing on disturbed ground, in gardens and in greenhouses. It is a 'pioneer' species, colonising bare ground, and it is often found growing where fires have scorched the soil.

This pompom of delicate threaded beads looks as though it is part of some elaborate jewellery or clothing accessory. The beads are formed in chains, but they will soon start to break off individually, because this is the fruiting head of an *Aspergillus* fungus and the beads are its spores, ready to fly.

Aspergillus fungal moulds occur everywhere. The spores can develop in almost any organic matter, and some will grow in nutrient-depleted sites such as damp walls, where they are a major component of mildews. However, they are particularly associated with wet hay, especially when it has been warmed up in order to dry it. Like all fungi, *Aspergillus* sends out tubular tendrils called hyphae, which penetrate their foodstuff, digesting it and absorbing the nutrients until they are ready to sprout new fruiting bodies. This type of fruiting body is called conidiophore, and the spores, in chains ready to break off, are conidia. They are produced without fertilization: there has been no combination of male and female genes, and each spore is genetically identical to the parent body that produced it.

The minute spores are airborne, and form part of a diverse aerial plankton. They are so small that humans breathe them in and out and usually do not notice them. If a spore settles in the lungs it is quickly removed by the body's immune system. Unfortunately, there are some individuals whose immunity has been compromised, either because of an inherited genetic factor or by deliberate drug suppression following transplant surgery. In these cases, the fungus can grow in the lungs, one of only a few species that thrive in the relatively high temperature (37 °C) of the human body. The condition, called aspergillosis, can be fatal. Another animal attacked by the fungus is the honeybee. Larvae infected with the fungus become blackened and hard after death, giving the disease its colloquial name of stonebrood.

RIGHT
This spore capsule of *Funaria hygrometrica* shows the peristome, the toothed ring protecting the mouth of the capsule, and some spores.

LEFT
Mosses grow on bare substrates like rock, walls and roofs, where they are not shaded out by other plants.

RIGHT
The bead-like spores of an *Aspergillus* fruiting body are ready to be wafted into the air.

LEFT
Although harmless to healthy people, *Aspergillus* can attack those with respiratory problems or in whom the immune system is weakened.

Two simple bulbous blobs, but the fissure down the length of each offers the promise of something hidden inside. The secret within is the male genetic line of the plant, because these are the anthers and what they hold is pollen.

The male sexual organs of a flowering plant are its stamens (sometimes called an androecium). Each comprises a thin stalk called the filament (coloured green here), and at the tip an anther (here coloured white). The anther produces pollen, which, depending on the species, is then transferred by wind or pollinator or some other process to fertilize the ovules in the female flower, often on another plant.

Pollen is microscopic and contains little except the male genetic material. Inside the anther, cells divide by a process called meiosis, or genetic reduction. During this phase the amount of genetic information, in the form of DNA molecules grouped into genes arranged on chromosomes, is halved, a situation known as the haploid condition.

A parallel genetic reduction occurs when ovules are produced in the ovaries of female flowers. When a haploid pollen grain fertilizes a haploid egg in a female flower, the full double (diploid) genetic complement is achieved and the resulting seed is ready to develop into a complete new plant.

As the pollen is created inside the anther, it reaches a point of maturity when the anther splits open. In animal-pollinated flowers this coincides with the availability of nectar to attract mainly flying insects, but also hummingbirds, bats and other creatures. Pollen brushing onto the bodies of these pollinators is then transferred to other flowers of the same plant species.

The number of stamens, and the corresponding number of ovaries, in a given flower have long been used as a means of classifying plants. In times past, there was some titillating controversy when this was couched in terms of so many gentlemen being in a bed of petals with so many ladies.

Arranged like clusters of walnuts in a greengrocer's display, these knobbly fruits look ripe for the picking. But these seeds will not drop heavily to the ground when they are ready, they will waft on the wind. Bracken spores are minute, the size of dust particles, and they will be carried many miles through the air.

Like mosses, ferns have an alternating series of generations. Spores germinate into a tiny growth of green cells called a prothallus, with male and female portions. Male sperm swim through leaf surface moisture to fertilize the female ovules, and a new fern plant grows from this. Unlike mosses, which produce specialised spore capsules (see page 195), the whole fern frond is a spore bearer, and its under surface is covered by pockets called sporangia, which contain the spores.

Bracken is one of the most widespread ferns in the world, occurring on every continent except Antarctica and producing thick dense stands, usually on slightly acid soil. Strangely, bracken generation from spores is very rare. Instead it spreads by sending out underground stems (rhizomes) from which new shoots sprout up. Spores seem to give rise to new plants only in fairly sterile conditions where they are not attacked by fungus, such as after fire has scorched the ground or in the mortar of decaying buildings.

Most bracken foliage is toxic to livestock and humans, but the young fronds are sometimes cooked and eaten in Japan, even though they are reported to be carcinogenic. The spores also contain carcinogenic chemicals, and there have been worries that people who regularly walk or work in bracken-dense forests and moors are at some slight risk. Despite these fears, and the fact that bracken is classed as a noxious weed in many countries, it remains an important and valuable feature in the landscape, providing shelter and food for many animals.

RIGHT
The anthers, bulb-like lobes extended on stalks, produce and release the plant's male sex cells – pollen.

LEFT
The chameleon plant is a native of Japan, China and southeast Asia, but is now planted in gardens worldwide for its pretty foliage and insect-attracting flowers.

RIGHT
On the underside of a bracken frond, the sporangia have split and peeled back, revealing the tiny spores now ready to be released into the air.

LEFT
Despite containing known carcinogens, young bracken fronds are cooked and eaten in some parts of East Asia.

GLOSSARY

Acetabulum
Socket that receives a bone, leg or antenna; also the hollow inside a sucker-cup. Named after its resemblance to a vinegar-cup.

Acetylcholine
A neurotransmitter produced at the end of nerve fibres when they activate a muscle.

Aedeagus
Hard penetrative organ of a male insect, analogous to the penis.

Alga
Simple plant; includes single-celled forms as well as pondweeds and seaweeds (plural *algae*).

Androecium
Male parts of a flower, comprising the stamens.

Animalcule
Any microscopic animal.

Antennifer
Socket which receives the basal segment of the antenna of an insect.

Arolium
Pad between the claws at the end of an insect leg.

Barbule
A small barb, more particularly one of the curved interlocking hairs on the barbs of a feather.

Binary fission
Division into two; simple means of reproduction where the body of a single-celled organism grows larger, then divides into two 'daughter' cells.

Bolus
Rounded mass or lump.

Calymma
Outer layer of cytoplasm covering the silica skeleton of a radiolarian.

Carbohydrate
Compound of carbon, hydrogen and oxygen, particularly sugars and starches.

Cartilage
Flexible skeletal tissue; in vertebrates the embryonic skeleton, most of which turns to bone, except in some fish.

Cephalothorax
Fused head and thorax segment in spiders, scorpions and other arachnids.

Cercaria
Final larval stage of trematode worms (plural *cercariae*).

Chemoreceptor
Sensory nerve ending that reacts to a chemical stimulus.

Chimaera
Mythical mixed animal, often made up of lion head, goat body and serpent tail; plant or animal made up of two genetically distinct tissues.

Chitin
Complex carbohydrate molecule laid down in overlapping sheets in the cuticle of insects and other animals.

Chlorophyll
Magnesium-containing green pigment in plants, used in photosynthesis.

Chromosome
Condensed aggregations of DNA, usually only visible during cell division.

Cilium
Minute hair attached to the outside of an animal cell (plural *cilia*).

Circumvallate
Surrounded by a ring or wall.

Collagen
A protein in fibrous connective tissue.

Commensal
Living together (originally eating at the same table); organism living on another without doing its host any harm.

Compound eye
Eye, possessed by insects and a few other animals, made up of many facets, each with its own lens and light receptor.

Conidium
Non-sexual (not male or female) fungal spore (plural *conidia*).

Cultivar
Cultivated form or variety of a plant.

Cysticercus
Larval form of certain tapeworms (plural *cysticerci*).

Cytoplasm
Semi-liquid or gel-like protein matrix inside a living cell.

Cytostome
Mouth-like opening of single-celled animal, through which food is injested.

Denticle
A small tooth or tooth-like structure.

Dentine
Hard calcareous compound of which teeth are composed.

Detritivore
Scavenger, something that feeds on detritus or debris.

Diatom
Single-celled alga, coated with hard silica shell in two halves.

Dicotyledon
Major group of (most) flowering plants, where seeds contain two embryo leaves rather than one.

Diploid
Having the full complement of chromosomes for a particular species, usually an even number because they occur in corresponding pairs.

DNA
Deoxyribonucleic acid. Convoluted double-helix molecule, found in the chromosomes of plants and animals, containing the genetic information of an organism.

Ectoplasm
Outer layer of cytoplasm of animal cell.

Endophallus
Flexible and inflatable part of insect genitalia through which sperm is injected.

Endoplasm
Inner portion of the cytoplasm of an animal cell.

Endosperm
Food store for the embryo in a plant seed.

Epidermis
Outer layer of skin.

Exine
Tough outer coat of a pollen grain.

Exoskeleton
Hard, usually jointed, outer shell of insects, scorpions, crabs and other animals.

Fallopian tube
Tube through which unfertilized eggs pass from an ovary to the womb.

Filiform
Thread-like, long and thin.

Flagellum
Long cilium, or hair, on an animal cell (plural *flagella*); also the long whip-like part of an antenna.

Foliate
Having leaves, or looking like a leaf.

Follicle
Small sac-like structure, particularly that around the root of a hair (originally meant a small windbag).

Frenulum
Curved hair on the hind wing of a moth uniting front and back wings during flight.

Fungiform
Mushroom-shaped.

Frustule
Two-piece silica shell of diatom, a single-celled alga.

Furcula
Forked springing organ of a springtail; also the wishbone of birds.

Galea
Flexible appendage, two of which together make up the sucking tongue of butterfly or moth (plural *galeae*).

Gall
Abnormal growth of a plant in response to an insect (or other small creature) feeding on or inside the tissue of the leaf, stem, bud etc.

Glucosinolates
Toxic carbohydrate substances found in uncooked cabbage leaves.

Guard cell
One of a pair of curved cells that form the mouth of a stoma, a plant breathing pore.

Gynaecophoric channel
Groove in a male trematode worm which houses the female.

Gynoecium
Female parts of a flower, comprising the pistil and ovaries.

Haemoglobin
Iron-containing red pigment in blood and body fluid of animals, used for transporting oxygen and carbon dioxide.

Haemolymph
Body fluid of insects.

Haltere
Vestigial hind wings in flies, reduced to small knobs and used as balancers to detect movements when flying.

Hammuli
Row of small hooks on the hind wing of bees, wasps and ants, uniting front and back wings during flight.

Haploid
Having only half the normal number of chromosomes for a particular species, one from each chromosome pair (usually only occurring in egg, and sperm or pollen).

Histamine
Active biological compound which accumulates during allergic reactions causing dilation of blood vessels.

Hydrophobic
Water-repellent.

Hyphae
Slender tubular filaments forming the nutrient-collecting but non-fruiting part of a fungus (singular *hypha*).

Infundibulum
Funnel-shaped cavity; in particular, part of the sucker on an octopus arm.

Keratin
Fibrous animal protein found in hair, wool, horn, nails, claws, hooves, scales and feathers.

Kinetosome
Pore from which one or more hair-like cilia sprout on the outer surface of a single-celled animal.

Labium
A lip, more particularly the lower (floor) portion of the insect mouth.

Lamella
A thin plate or scale, in particular the flattened terminal segments of certain insect antennae (plural *lamellae*).

Maxilla
One of the mouthparts of an insect or other arthropod used with the mandibles for chewing (plural *maxillae*).

Melanin
Ubiquitous dark pigment found throughout the plant and animal kingdoms.

Mellitin
The major constituent of bee venom.

Metamorphosis
Changing from one form to another, especially the transformation from larva to adult insect.

Microfibril
Minute or sub-cellular fibre.

Micrograph
Image collected by microscope; may or may not be a photograph, depending on whether light was used.

Microtichium
Minute hair on the wing of an insect (plural *microtrichia*).

Miracidium
First larval stage of a trematode worm (plural *miracidia*).

Mitochondrion
Cell organelle where nutrients are metabolized into energy (plural *mitochondria*).

Monocotyledon
Major group of flowering plants (mainly grasses, sedges, orchids, lilies and irises) where seeds contain one embryo leaf rather than two.

Mystacial
To do with moustaches.

Notochord
Rod of cartilage forming the embryo basis of the spinal cord in vertebrates.

Oesophagus
The gullet; the tube that takes food from the mouth to the stomach.

Ommatidium
Single facet of a compound eye (plural *ommatidia*).

Oocyte
Female germ cell which awaits fertilisation from sperm or pollen.

Organelle
Identifiable specialised structure inside a single plant or animal cell which acts as an organ.

Osmosis
Diffusion of a substance through a membrane.

Papilla
Small nipple-like mound (plural *papillae*).

Paramere
Lobe (usually paired) of the penetrative part of male insect genitalia.

Parasite
Organism that lives off another by feeding on it, or by stealing nutrients from it, but which does not kill its host.

Parasitoid
Organism that feeds on or inside another and which eventually kills it.

Pedicel
A stalk, more especially the second segment of an insect antenna between the scape and the flagellum.

Pheromone
Chemical scent released by an animal which signals to others or influences their behaviour.

Phloem
Tubular vessels through plant stems and leaves transporting liquid nutrients.

Photosynthesis
Complex manufacture of carbohydrates by the pigment chlorophyll in green plants, combining water and carbon dioxide using the energy in sunlight.

Phytoplankton
Oceanic single-celled algae.

Placoid
Resembling plates, of certain shark, ray and other fish scales that are not overlapping.

Polymer
Long molecule made up of repeated shorter units.

Polysaccharide
Molecule, such as starch, made of repeated sugar units.

Polytene
Grossly replicated strands of DNA formed into giant chromosomes in certain fly maggots.

Prehensile
Capable of gripping or grasping.

Proboscis
Snout, nose or trunk; sucking mouthparts of some insects.

Proglottid
Detachable segment of a tapeworm.

Proleg
Leg on an abdominal segment of an insect larva, distinguished from a true thoracic leg.

Prothallus
Small leaf-like growth in the life cycle of ferns, mosses etc, where male and female cells combine to create spore-bearing structures.

Protozoan
Single-celled microscopic organism, including amoebas.

Pseudopodia
Blob-like extensions of the soft body of single-celled animals, which move it along or reach out, envelop and then absorb food particles.

Pulvillus
Cushion-like pad between the claws on an insect leg.

Pupa
A chrysalis; the hard shell covering an immature insect as it transforms from larva to adult.

Puparium
Last larval skin, particularly of a fly maggot, inflated and hardened to protect the pupa developing inside.

Rachis
A spine; the axis of a feather, inflorescence or stem of leaflets.

Radiolarian
Single-celled animals containing circular or spherical skeletal frame made of silica.

Radula
The rasping ribbon of a mollusc tongue.

Retina
The light-sensitive layer in an eye.

Retinaculum
Curved hook on the underside of some butterfly and moth forewings which holds the wings together in flight.

Rhabdome
Central light-sensitive core beneath each facet of a compound eye.

Rhizome
Un underground stem from which roots and shoots emerge.

Sanguivorous
Feeding on blood.

Scape
The first, often much longer, segment of an insect antenna.

Sensillum
A small sense organ on the body, especially the antennae, of an insect (plural *sensillae*).

Serotonin
Neurotransmitter substance released from nerve cells.

Seta
Hair-like structure on the surface of an organism (plural *setae*).

Shagreen
Originally a granular leather made from horse or donkey skin, now usually shark or ray skin; shagreened – covered in shagreen or dulled by using shagreen as sandpaper.

Sinigrin
Toxin in cabbage leaves, harvested by large white butterfly caterpillars to make them distasteful to predators.

Sinus
Air- or fluid-filled void.

Spermatheca
Organ in female insect body where sperm is stored after mating.

Spidroin
Major component of spider silk.

Spigot
Closable hole through which passing liquid can be controlled.

Spinnerets
The spinning organs of spiders.

Spiracle
A breathing hole, such as that through the exoskeleton of an insect.

Sporophyte
The spore-bearing part of a fern or moss.

Sporopollenin
Major component of outer skin of pollen grains.

Stoma
Breathing pore in a plant (plural *stomata*).

Stylet
Small pointed object, more particularly one of the tubular arrangements of piercing mouthparts in some insects.

Symbiosis
Living together in mutual harmony, both benefiting (adjective *symbiotic*).

Tarsomere
Segment of the tarsus (foot) of an insect or other animal.

Tenent setae
Small spatula-shaped hairs on an animal foot used for holding on to flat surfaces.

Terpene
Hydrocarbon found in the essential oils of certain plants.

Testa
A hard shell or seed coat.

Trachea
Internal breathing tube.

Trichium
Hair-like structure on the surface of an organism (plural *trichia*).

Trichome
A plant hair.

Ungulate
Walking on hoofs, or on tiptoe.

Van der Waals force
Week attractive forces between molecules.

Vibrissa
A bristle sensitive to touch (plural *vibrissae*).

Villus
Finger- or frond-like extensions.

Xylem
Woody tubular vessels in plant stems and tree trunks that conduct water from roots to leaves.

PICTURE CREDITS

BIN TRAVELER FORM

Cut By _Josias Fandu of_ Qty _25_ Date _7/5_

Scanned By _[signature]_ Qty _____ Date _3-5-24_

Scanned Batch IDs
_1114193.32_ / _1114145.47_ . / _11140/300_ .

Notes/Exceptions

_____

_____

_____

_____

_____

_____